ISBN 978-1-330-07572-2
PIBN 10019614

English
Français
Deutsche
Italiano
Español
Português

www.forgottenbooks.com

Mythology Photography **Fiction**
Fishing Christianity **Art** Cooking
Essays Buddhism Freemasonry
Medicine **Biology** Music **Ancient
Egypt** Evolution Carpentry Physics
Dance Geology **Mathematics** Fitness
Shakespeare **Folklore** Yoga Marketing
Confidence Immortality Biographies
Poetry **Psychology** Witchcraft
Electronics Chemistry History **Law**
Accounting **Philosophy** Anthropology
Alchemy Drama Quantum Mechanics
Atheism Sexual Health **Ancient History**
Entrepreneurship Languages Sport
Paleontology Needlework Islam
Metaphysics Investment Archaeology
Parenting Statistics Criminology
Motivational

IS THE

COPERNICAN SYSTEM

—OF—

ASTRONOMY TRUE?

BY W. S. CASSEDY,

Of Pittsburgh, Pa.

KITTANNING, PA.

STANDARD PUBLISHING CO.

1888.

CONTENTS.

CHAPTER I.

Astronomers say, in speaking of the distance of the sun,
"The determination of this distance is therefore one of
"the capital *problems* of astronomy, as well as one of the
"most difficult, to the solution of which both ancient and
"modern astronomers have devoted many efforts."

The importance of this knowledge, as a means of determining the distance of all other heavenly bodies, may be
judged of, and the uncertainties of it, by the following
quotations from "Newcomb's Popular Astronomy"; he
says : "Recourse must be had to methods of observation
"subject to many sources of error." Again, "The measure-
"ments of the heavens involves two separate operations.
"The one consists in the determination of the distance be-
"tween the earth and the sun, which is made to depend on
"the solar parallax, or the angle which the semi-diameter
"of the earth subtends as seen from the *sun*, and which is
"the *unit* of distance in celestial measurements. The
"other consists in the determination of the distances of
"the stars and planets in terms of this unit which gives
"what we may call the proportions of the universe.

"Knowing this proportion, we can determine all the
"distances, when the length of our unit, or the distance of
"the sun is known, but *not before.*"

The planets which are nearest the earth are regarded
by all astronomers much nearer us than is the sun.

The reliability of the method by parallax for finding
the distances of the planets may be judged of by the following quotation from "Newcomb's Popular Astronomy";
he says :

"The earth is so small in comparison with the distances "of the planets that the parallax in question almost eludes "measurement, except in the case of those planets which "are nearest the earth, and even then it is so minute, that "its accurate determination is one of the most difficult "*problems* of modern astronomy."

The planet Venus—one nearest us—is estimated to be 33,000,000 miles nearer the earth than is the sun. The parallax of this comparatively near planet "is so minute "that its accurate determination is one of the most difficult "problems of modern astronomy."

As parallax is the method by which the sun's vastly greater distance is' to be found, is it possible by this method to have any knowledge, even by approximation, of the *real* distance of the sun ?

Before proceeding further with this method and its different applications in finding the sun's distance, let the enquiry be directed as to the reliability of the method for finding the distances of the stars.

Parallax is defined to be "the difference between the direction of a body as seen from two different points," being a mathematical process employed in the solution of a triangle.

The difficulty of finding the distance of a star by this method may be estimated, and, indeed, is found to be insuperable, from the fact that if the position of a star be observed, say, for example, in June, and again in December, it will be found in the same relative position in the latter month as in the former.

Yet, according to the Copernican system of astronomy, in the interval between these observations the earth has progressed in its orbit with a velocity of 67,000 miles per hour, and has travelled a distance of 190 millions of miles measured in a straight line across this orbit, and about 285 millions of miles measured around its curve, and, notwithstanding this startling circumstance, not the slightest angle can be observed between these periods of observation on attempting the parallax of a star.

What is, in this connection, worth observation as a remarkable coincidence, is the fact that this absence of apparent displacement of the observed star would be observed did the earth not progress in the assigned orbit.

The unreliability of parallax to find the distance of any star may be judged of by the following conclusion of Sir John Herschel, he says :

"It might be naturally enough expected by this enlarge-"ment of our base to the vast diameter of the earth's orbit, "the next step in our survey would be made at a great "advantage; that our change of station from side to side "of it would produce a perceptible and measurable amount "of annual parallax in the stars, and that by its means we "should come to a knowledge of their distances. But after "exhausting every refinement of observation, astronomers "have been unable to come to any positive and coincident "conclusion upon this head; and it seems, therefore, *dem-* "*onstrated* that the amount of such parallax, even for the "nearest fixed star which has been examined with the re-"quisite attention, remains still mixed up, and concealed "among the errors incidental to all astronomical determi-"nations. Now such is the nicety to which these have "been carried, that did the quantity in question amount to "a single second,* that is, did the radius of the earth's or-"bit subtend at the nearest fixed star that minute angle, it "could not possibly have escaped detection and universal "recognition." *Herschel's Astronomy, Title "Fixed Stars."*

Either one of two conclusions may, possibly, be deduced from the last foregoing quotation, either of which would account for the failure of these attempts to find the distance of any star. One of these may be—as is generally supposed to be the case—because the distances of all the stars are so great that *no* angle is observable; the other is

---

* Namely, the width of a hair,

that no star has been observed really from two *different* positions, which last would be the case did the earth not progress in reality in the theoretical orbit assigned it.

Whatever may be the cause, the foregoing quotation shows that the method by parallax fails to find the distance of any star.

To show still further that the method is not reliable for finding the distances of the planets, we quote from M. Biot's work, he says :

"The parallax of *any* one planet, *if it were known*, would "give immediately its absolute distance from the earth; "but these parallaxes are in general too small to be accu-"rately observed." Sec. 513.

In Sec. 514, he says of the planet Mars : "Unhappily "the parallax of Mars is so small that the errors of obser-"vation may have a very considerable influence upon the "result, and consequently upon the values of the distances "determined by means of it. If, for example, in deter-"mining the horizontal parallax we commit an error of "2", or $\frac{1}{5}$ of its total value, there would be an error of $\frac{1}{5}$ in "the estimate of the distance.

"Yet we *never* can be secure against so small an error "in observations so delicate; for 2" does not amount to "*half* the breadth of the *wire* used in the micrometer."

Judging by this admission, it appears that it cannot be expected that parallax would be any more successful in finding the distance of the sun, as it is supposed to be a more distant body than the nearest planet.

Indeed, the uncertainties of the direct solar parallax were found to be so great, and the method so unreliable, that it is abandoned, as we are informed by Prof. Newcomb, who says the modern method consists, "not in meas-"uring the parallax directly, because this cannot even now "be done with *any* accuracy, but in measuring the paral-"lax of one of the planets Venus and Mars when nearest "the earth."

How reliably the parallax of Venus and Mars has been found, may be inferred from the last preceding quotation from M. Biot, who says of observations so delicate as these "we *never* can be secure."

The following quotation contains Sir John Herschel's views in regard to direct and indirect parallax to find the the distance of the sun :

"The diameter of the earth has served us as the base of "a triangle in the trigonometrical survey of our system, "by which to calculate the distance of the sun ; but the "extreme minuteness of the sun's parallax renders the cal- "culation from this 'ill-conditioned' triangle so delicate "that nothing but the fortunate combination of favorable "circumstances afforded by the transit of Venus could "render its results *even tolerably* worthy of reliance."

The foregoing amounts to another astronomical admission that direct parallax has failed to find the distance of the sun.

It also amounts to the admission that the indirect parallax on the transit of Venus is only "tolerably worthy of reliance," owing to "the fortunate combination of favorable circumstances" which this transit affords.

This "fortunate combination" does not appear as a matter of congratulation in view of what M. Biot says in regard to the parallax of the planets, namely: "The parallax of any one planet *if* it were *known*," &c., which has been before quoted.

In point of fact, what does "the fortunate combination of favorable circumstances" amount to in finding the distance of the sun on the transit of Venus, unless the distance of Venus were known ; and this cannot be known, for as Prof. Newcomb informs us in a previous quotation, in speaking of the distance of the sun, "this is the unit of distance in celestial measurements," and that, "the dis- "tance of stars and *planets* are determined in terms of this "unit. We can determine all the distances, *when* the "length of our unit, or the distance of the sun is known, "but *not before.* From this point of view, the distance of Venus cannot be known until the distance of the sun is known. What then becomes of the parallax of Venus? What reliance can be placed on the sun's distance found by parallax on the transit of Venus?

It would appear from the following quotation that some reliance is placed on Kepler's third law for finding the distance of the sun ; this law is as follows, namely :

"The square of the time of revolution of each planet is "proportional to the cube of its mean distance from the "sun."

In speaking of this law, Prof. Newcomb, in his Popular Astronomy, says :

"The ratio between the distances of the planet (Venus) "and the sun is known with great exactness by Kepler's "third law, from which, knowing the differences of the "parallaxes, the distance of each body can be determined."

Not the real distance of the sun, but the *proportional* distance between Venus and the sun, is what is determined by Kepler's third law. For if this third law finds the *real* distance of the sun—which is implied in the last quotation—why rely on parallax at all, of any sort, for finding the distance of the sun? In such case, this third law of Kepler enables us to know the distance.

It either finds the distance of the earth—as one of the asserted planets—from the sun, or it does not.

If it does this, taking the parallax on the transit of Venus is unnecessary for this purpose.

If it does not find such distance, then "the proportions of the system," as determined by such third law, cannot be relied on for finding the real parallaxes of Venus and the sun, because this law fails to find the real distance of another planet, namely, the earth ; yet, by the third law of Kepler, the distances of *all* the planets are pretended to be determined.

Besides, is it not probable that this strange relation of the cube of the distance being proportional to the square of the time of revolution, is rather an instance of the many curious and inexplicable relations of figures and quantities, to which an ingenious mind may discover a plausible but not, necessarily, a correct signification.

Because, if this third law be a discovery of a relation existing betwixt velocity and distance, it is difficult to see why the distance of the moon might not be found by it, and, indeed, the distance to the centre of the earth, were this distance unknown; for a body on the earth's surface revolves about the earth's centre in a regular time, if the earth rotates, just as the planets do about the sun.

But to return to a further consideration of parallaxes.

The parallax on the transit of Venus was attempted by Halley, in A. D. 1661, and again in 1669, and no reliable result obtained, mainly, it was thought, because of an elongation or neck observed between the limb of the sun and the planet, called by astronomers "the black drop."

The result of this—as stated by Prof. Newcomb—was "an uncertainty, sometimes amounting to nearly a minute, "in observations which were expected to be correct within "a single second."

In consequence of the presence of this "black drop"— this Banquo's ghost at the intellectual astronomical banquet—but little reliance was placed on the parallax found on these occasions.

In order to show that the parallax sought was not found to be any more reliable on the transit of Venus, in A. D. 1876, we quote from Newcomb's Astronomy, as follows:

"The fact is, that although a century ago a transit of "Venus afforded the *most* accurate way of obtaining the "distance of the sun, yet the great advances made during "the present generation in the art of observing, and the "application of scientific methods have led to other means

"of greater accuracy than these old observations, it is re-
"markable that while nearly every class of observations is
"now made with a precision which the astronomers of a
"century ago never thought possible, yet this particular
"observation of the interior contact of a planet with the
"limb of the sun has *never* been made with anything like
"the accuracy which Halley himself thought he attained
"in his observation of the transit of Mercury two cen-
"turies ago."

The parallax of Venus A. D. 1876, then, is no more
reliable for finding the distance of the sun than that taken
two centuries ago, which latter was expected to be so accurate as to be within a single second, while, in fact, it
varied from this to the extent of a minute, thus not being
within 59 seconds of the reliability expected, that is to
say, giving only the $\frac{1}{60}$ of the accuracy in results expected.

Yet M. Biot says—as before quoted—"If, for example,
"in determining the horizontal parallax (of Mars) we
"commit an error of *two* seconds, or $\frac{1}{5}$ its total value, there
"would be an error of $\frac{1}{5}$ in the estimate of the distance,"
in such case missing its mark nearly thirty millions of
miles.

While in Halley's attempt to take the parallax of Venus there was an error of 59 seconds, yet Prof. Newcomb
says, virtually, that the parallax has "*never* been made
with anything like the accuracy" of Halley's effort "two
centuries ago."

From these recitals and results, parallax, as a means of
finding the distance of the sun, may, confidently, be pro-
nounced a failure.

It is practically conceded to be so, by astronomers themselves, as appears from the following extract taken from
"Newcomb's Popular Astronomy," namely:

"The method of obtaining the astronomical unit (name-
"ly, the distance of the sun,) which we have described
"rests entirely upon the measure of parallax, an angle

"which hardly ever exceeds 20″ and which, therefore, is "*exceedingly* difficult to measure with the necessary ac-"curacy.

"If there were no other way than this of determining "the sun's distance, we might despair of being sure of it "within 200,000 miles.*

"But the refined investigations of modern science have "brought to light other methods, by at least two of which "we may *hope ultimately* to attain a greater degree of ac-"curacy than we can by measuring parallaxes.

"Of these two, one depends on the gravitating force of "the sun upon the moon, and the other upon the velocity "of light."

Taking one view of this admission, it might be inferred that no parallax can determine the distance of the sun ; in another view, its possible accuracy may be implied, say; for example, within a distance of 200,000 miles.

The vast, general capacity for inaccuracy in the method by parallax implied in these figures, namely, in parallax missing its mark to the extent of a distance equal to 25 times the diameter of the earth, almost as far as the moon when nearest the earth, would naturally induct the mind to the conclusion that this method was capable of still greater achievements in this uncertain line, to the extent of not finding the distance of the sun even approximately.

If the extent or limitation of the inaccuracy were un-known, it, as a consequence, would not be a measurable quantity, hence neither 200,000 miles nor any other dis-tance could be the measure of such unknown inaccuracy, for where the inaccuracy cannot be measured how is the *extent* of it to be made known ?

If such inaccuracy were a measurable quantity, then there need be *no* inaccuracy, for such quantity might be

---

* That is to say, the same method which fails to find the distance of Venus with "*certainty*" (as just quoted) finds, within 200,000 miles, the distance of the sun estimated by astronomers to be 60 millions of miles *further* from the earth than is Venus.

added or subtracted, as the case might require, and thus parallax would become reliable. Hence, it may be inferred that the 200,000 miles estimate of the deficiency or error of the parallax is but a conjecture.

For were the actual distance of the sun not found by parallax, it must be because there must be some defect in the method so applied, or in the instrumentalities employed, or in some disturbing and immeasurable fact, and were such the case it would follow that *no* reliance whatever, approximately or otherwise, could rationally be placed on the result.

Some of the difficulties encountered in making the parallax, when the telescope is used, may be judged of when astromomers tell us, as they do, "that the lens will "not bring all the light to the same focus, and presents a "confused image of the object to the observer." Another source of error may be traced to secondary aberration, which arises from a "property in glass as glass."

Again, "irradiation may arise from a number of causes : "imperfection of the eye, imperfection of the lenses of the "telescope, and the softening effects of the atmosphere, &c." Thus indirect parallax appears unreliable.

But why does not the *"annual* parallax" find the distance of the sun ?

The distances of the stars have been attempted to be found by this method without success, which failure is attributed to the great distances of the stars, but we are told by astronomers that "the sun is two million radii of the earth's orbit" *nearer* the earth than is the nearest star.

If the Copernican system of astronomy be true, why resort to the *indirect* parallax in order to find the distance of the sun ?

The earth's orbit furnishes the *same* base line from which the sun's distance may be measured, as that from which the distances of the stars have been sought.

The straight line across the earth's orbit is 190 millions of miles in length. The smallness of the earth's diameter is given as the *reason* why the distance of the sun cannot be found by parallax.

The base line of the earth's orbit is more than 23,000 times longer than the diameter of the earth, which is nearly equal to the difference between a foot in length 4½ miles.

According to the Copernican system the sun is stationary, and the earth moves between the tropics.

Here may be found the "two different points" of observation required by the method by parallax, and the "difference in direction" required may be found along a line 190 millions of miles in length. The two observations could be make six months apart, like in the annual stellar parallax.

The size of the angle, in comparison with that subtended were the earth's diameter employed to form a base line, would be in the proportion of 23,000 to 1.

Yet the distance of the sun has not been found, notwithstanding these favorable conditions; for the "ill conditioned triangle," complained of by Sir John Herschel, when the earth's diameter served as a base line with which to find the distance of the sun by parallax, cannot here be objected to.

Why is the distance of the sun not found from the diameter of the earth's orbit ?

It must be either,

1. Because the orbit only exists "in the mind's ideal shape of such," in which case the necessary "two different points," namely, along the diameter of the earth's orbit, have no existence beyond "the intendment of the mind," and were such the case the Copernican system of astronomy would necessarily be a myth, or,

2. Because there is some defect in the method by parallax when applied to bodies so distant as the heavenly

bodies, or else such other favorable conditions wanting, as to render the method inapplicable to the end sought by it.

If the "two different points" required for the observations have no existence, what becomes of "the Copernican System of Astronomy ?"

If the "two different points" really have an existence—and it be.conceded that the base line is long enough—then either the method, or some necessary instrumentalities, must be inefficient to find the distance sought.

Should such be the case, what reliance can be placed on any parallax having for its object the finding of the distance of any heavenly body.

If the orbit exists, and the angle found from it be large enough, and the method by parallax be competent to find the distance of a heavenly body, why is the distance of the sun yet "a problem," as astronomers say it is?

SECTION II.　THE VELOCITY OF LIGHT.

When Jupiter is farthest from the earth, the light from its satellites is observed about 17 minutes later than when the planet is nearest the earth.

From this observed phenomenon, astronomers have come to the conclusion that it requires 17 minutes longer time for light to travel from the satellite to the earth in the former case than in the latter.

This astronomical conclusion will be seen by the following quotation from M. Biot's work.　He says, Sec. 461 :

"The most simple explanation which suggests itself is "to *suppose* that the light of the sun, reflected by these "small bodies, is not transmitted instantaneously to the "earth, but that it occupies a small interval of time in "traversing the earth's orbit."

It will be perceived that this conclusion is based on the theory, now superannuated, that light is a substance and that it travels.

In order to determine the distance of the sun by means of the motion of light from a satellite of Jupiter, which was proposed, it became of the first importance to find the velocity of light.

One of the most approved methods employed for this purpose may be understood by the following—although partial—account of it, taken from Newcomb's "Popular Astronomy :"

"We have a wheel with spokes, extending from the cir-
"cumference, the distance between them being equal to
"their breadth.

"The wheel is placed in front of the lantern so that the
"light from the latter has to pass between the spokes of
"the wheel in order to reach a distant mirror. The eye
"of the observer is between the spokes. * * * *

"Now suppose he turns the wheel, still keeping his eye
"on the same point. Then each spoke cutting off the
"light of the lantern as it passes, there will be a succession
"of flashes of light which will pass through the spokes,
"*travel* to the mirror, and thence be reflected back to the
"wheel."

On this principle it was determined that light travels 180,000 miles per second.

It will be seen that the phenomena are observed by the naked eye. The conclusion reached as to the velocity of the flashes, and thence the velocity of light, are deduced from the impression made on the retina. Can this be relied on for determining the velocity of light, even though it were admitted that it does travel ?

In order to show how light acts on the retina we quote from Frank Leslie's (May No., 1881,) Popular Monthly Magazine, N.Y., part of an article entitled "Optical Illusions," by Conrad W. Cooke. He says:

"Every boy is familiar with the experiment of making
"a ring of fire in the air by swinging round the red-hot
"end of a burning stick. The luminous ring so formed is

"obviously an illusion, for it is clear that the light from
"the incandescent point can come only from one position
"in its path at any one time.   It cannot be at the same
"instant at both ends of the diameter of the circle, and yet
"the eye can detect no break in the continuity of its path.
"This experiment is a simple and characteristic illustra-
"tion of a large class of optical illusions, which result from
"a very necessary property of vision, which is called the
" 'persistance of visual impressions on the retina,' that is
"to say, an object placed before the eye, and suddenly re-
"moved, is seen for a certain appreciable time after its
"removal. * * * * * The time that this impression
"lasts has been variously estimated from the sixth to the
"eleventh part of a second.   The explanation, therefore,
"of a luminous ring formed by a lighted stick is that the
"impression made by it at any one point of its course re-
"mains on the retina until it again reaches that point."

Would not the same law apply to the flashes observed
between the spokes?   If so, what reliance can be placed
on it for determining the velocity of light? yet this method
of finding the velocity of light is the one most approved
by astronomers, being in principle Fazean's method.

May not the philosophy of the method be that the
spokes intercept the vision, and that the light on the mirror does not travel to the vision, but merely that the
vision takes cognizance of the fact that it is on the mirror
when observed between the spokes ; for example, say we
observe the moon, close our eyes and in less than the 20th
part of a second open them again, and again see the moon,
it could not be said in such case that the light of the moon
had travelled to the eye in the 20th part of a second.

The experiments of modern scientists, and their conclusions from them, are adverse to the theory that light
travels.   Prof. Tyndall says :

"The most refined and demonstrative experiments has
"led philosophers to the conclusion that space is occupied

"by a substance almost infinitely elastic through which
"the *pulses* of light make their way.   The luminous ether
"fills stellar space, it makes the universe a whole, and
"renders possible the intercommunication of light and en-
"ergy between star and star.   But the subtle substance
"penetrates further; it surrounds the very atoms of solid
"and liquid substances; the ether is *everywhere* and its
"*pulsation* is light."

If the undulatory or pulsatory theory of light be
the correct one, and the ether be "everywhere and its pulsation is light," and "this is the theory now universally
received," then the last wave pulsation or undulation, that
is to say, the one nearest the eye, would make the impression, causing the phenomenon, light, to be made manifest
to the visual organ, thence to the consciousness.

A homely illustration may serve to convey an idea of
the action of light on the theory now "universally received."   Let a common walking stick represent the ether
which is "everywhere;" were its end thrust against our
person, such motion might serve in some comprehensible
sense to the pulsation of the ether.   In the case supposed,
only that end of the stick nearest us would touch us, although the stick considered as a mass or whole, and in a
sense the atoms composing it, would move by the pulsation
or thrust, yet the sense of touch would be made alive only
by the particles of the stick nearest us.   So, in accordance
with the pulsatory theory, we may suppose that the organ
of vision is made conscious of the presence of light by
some more remote, but unknown agency, yet acting on
the visual organ by the immediate agency of the ether,
which is everywhere, coming in contact with it by its pulsation, just as the sense of touch is made alive by those
atoms of the stick which touch us, although the remote
cause of this sensation was the thrust given to the wooden
mass,

If this conclusion be deducable from the pulsatory theory, it cannot be said that the light of the satellite travels across the earth's orbit—in the sense, say for example, that a train of cars travels on a railway—and, after the lapse of 17 minutes arrives at the eye as a depot.

By the method before detailed, it is found that light travels 180,000 miles per second. It is observed that the light from a satellite of Jupiter is seen 17 minutes later when that planet is furthest from the earth.

According to the theory for finding the distance of the sun by the velocity of light, it requires 17 minutes for light to cross the earth's orbit, and one-half this time to cross one-half of this orbit, at the centre of which astronomers place the sun, hence light would require 8½ minutes to come from the sun to the earth. 8½ minutes contain 510 seconds. Light travels 180,000 miles in one second, and $510 \times 180,000 = 91,800,000$, being the distance of the sun from the earth in miles.

By this method, the existence of the orbit is assumed as a necessary part of the "Copernican System." Its dimensions are ascertained by "the proportions of the system"—in which the earth is included as one of the planets —according to Kepler's 3rd law.

Did the orbit really exist, and were its circumference known, as pretended, it were as easy to know the distance of the sun, placed in the centre of it, as to find the length of the spoke of a wheel, did we know the circumference of such wheel.

The dimensions of the orbit are either known, or they are not known.

If known, the velocity of light, in order to find the distance of the sun, is unnecessary, because there could be no difficulty in finding the radius of such orbit at the centre of which the sun would be found.

If the dimensions of the earth's orbit be unknown, how

long a time will be required for light to travel across an *imaginary* orbit?

The cause of the appearance of light from Jupiter satellite 17 minutes later, when Jupiter is furthest from the earth, may be caused by the different *angle* at which Jupiter and his satellite would probably be seen by an observer on the earth after the lapse of about six years, such being the difference in times of such observations, and after Jupiter had progressed in his orbit the distance of 1482 millions of miles, which it does in six years, according to calculations of astronomers.

It would be little short of a miracle did the light from the satellite appear at the same time and at the same angle, after it had progressed with its primary 1482 millions of miles.

Sir John Herschel says there would be a displacement of the stars caused by the *angle* at which they would be seen from the earth as it progressed in its orbit. And if so, why would not the satellites of Jupiter and their emerging light be seen at a different angle, after the lapse of six years, caused by their own proper motion as satellites, together with the motion in common with their primary.

This should be expected whether the earth moved in an orbit or not.

Herschel, in speaking of two stars on a line with each other, says :

"Supposing them to lie at a great distance, one behind "the other, and to appear only by casual juxtaposition "nearly in the same line, it is evident that *any* motion of "the earth must subtend different angles at the two stars "juxtaposed, and must therefore produce different paral- "lactic displacements of them on the surface of the heav- "ens regarded as infinitely distant." *Herschel's Astronomy,* "*Fixed Stars.*"

2

This observation seems to establish the fact that where two bodies are observed in the heavens, situate with reference to each other as stated, motion of the observer, or, of course, of the bodies observed, changes the angle of vision, which necessarily would change the apparent position and would probably change the appearance of the observed bodies.

In the case of a satellite of Jupiter, the effect of the different angle of vision at which the satellite would be observed after the lapse of six years, would, almost necessarily, be that light, at the terminus of an eclipse of one of these satellites, would be seen later at one of these periods than it would be at another, for it will be conceded that the position of the satellite in space with reference to the position of the observer would necessarily have an effect on the time at which the satellite and its emerging light would appear to the observer.

In this view, the 17 minutes later, when the light from the satellite is observed, is an optical phenomenon as an effect of the different angle of vision at which the satellite is observed than that at which it were seen six years previously, hence independent of the travelling theory of light.

In other words, the light from the satellite is visible later because the satellite itself comes on the line of vision of the observer later at one time than it does at another.

As a consequence the organ of vision cannot observe the light until the satellite, and the light which emerges from it, becomes elevated to a line with the vision, which line—it is supposed, for the reasons given—has different elevations at the different periods of observation of the planet and satellite.

This idea may be illustrated, perhaps, by supposing that the earth were moved horizontally north, say, for example, some millions of miles, leaving the moon behind, and continuing to revolve in its present orbit.

The moon at 40° N., when full, appears not many degrees from the vertical in December. In June it does not appear far from 30° above the southern horizon.

From the position where the earth has just been placed in the imagination, the moon would, in all probability, be seen sooner in December than it would be six months after or before, namely, in June, and were the velocity of light from our satellite determined by the longer time required for its light to reach the observer on the earth in June than in December, it would be conceded that such calculation would fail to give correct results, as these differing phenomenon would depend wholly on the difference in the angle of vision. For our satellite and its light might appear later at one time than at another, as an effect of such different angle, hence the velocity of light could not be computed by these phenomena.

On the whole, the conclusion appears rational that the distance of the sun cannot be found by the velocity of light for the following reasons :

1. If the ether be everywhere, and its pulsation be light, we may rationally suppose that it pulsates at the lantern and the mirror, in the experiment mentioned made to find the velocity of light.

2. If the pulsation of the ether occurs *at* the illuminated body, whose light is reflected, the pulsation, and consequently the light, may, possibly, be propagated *instantly,* without the lapse of *any* interval of time, just as the thrust of the stick—in the illustration before given—would be propagated and made manifest at each end of the stick at the *same* instant of time.

3. The most careful and thorough modern experiments made by the ablest scientists, in connection with light, have led to the conclusion, "now universally received," that light is not a body, is not matter at all, but on the contrary is merely a phenomenon as sound is, without

knowing—as is known in the case of sound—the cause of the phenomenon.

4. Even though light were admitted to be a body, and that it travelled, its velocity would be unreliable in consequence of the "persistence of vision" which produces an optical illusion, in each experiment made to ascertain its velocity, as may be concluded from the extract made on the subject of "optical illusions."

From the nature of the medium whose velocity is sought, and from the admitted uncertainties in the experiments made as to the *extent* of the optical illusions produced by light it may be rationally concluded that such extent is not a measurable or otherwise known quantity.

SECTION III.   LEVERRIER'S METHODS FOR FINDING THE DISTANCE OF THE SUN.

Reliance is placed, by some astronomers, on a method of Leverrier's for finding the distance of the sun, which may be understood by the following quotations from Newcomb's Popular Astronomy :

"Strictly speaking, the moon does not revolve around "the earth any more than the earth around the moon ; "but by the principle of action and reaction, both move "around their common centre of gravity. The earth be-"ing 80 times as heavy as the moon this centre is situated "within the former about three-fourths of the way from "its centre to its surface."

We quote further, as follows :

"It is known from the theory of gravitation that the "earth, in consequence of the attraction of the moon, de-"scribes a. small monthly orbit around the common centre "of gravity of these two bodies, corresponding to the

"monthly revolution of the moon around the earth, or to
"speak with more precision, around the same common
"centre of gravity. If we know the mass (or weight) of
"the moon relatively to that of the earth and her distance,
"we can thus calculate the radius of the little orbit re-
"ferred to. In round numbers it is 3,000 miles."

From this basis, Leverrier's method for finding the distance of the sun is thus explained :

"This monthly oscillation of the earth will cause a cor-
"responding oscillation in the, longitude of the sun, and
"by measuring its *apparent* amount, we can tell how far
"the sun must be placed to make this amount correspond
"to, say, 3,000 miles. Leverrier's found this oscillation in
"arc to be 6″.50. From this he concluded the solar parallax to be 8′. 9″5″."

How could Leverrier have known the *value* of this appearance on the sun, unless he knew the real size of that
longitude where such appearance were manifested ?

For the appearance would have been proportional to
the *real* size of such longitude; and the real size of it, by
the requirements of the law of gravitation,—in order to
harmonize with the proportions of the solar system—would
depend upon the distance of the sun, which is the very
problem pretended to be solved by the *appearance* of the
earth's oscillation in a longitude of the sun.

In order to show the necessity of this knowledge in order to find the distance sought by this method, suppose
two spheres—one, twice the diameter of the other—to be
placed at the *same* distance from an observer, and the oscillation of a pendulum, at the place of observation, should,
for example, have such appearance on the smaller sphere
that the chord of the arc of its oscillation were apparently
equal in length to the apparent diameter of such smaller
sphere, while such chord would appear but equal to the
radius of the larger sphere, and such different appearances
would result, in the case supposed.

While the chord of the arc of oscillation would really be of the same length on both spheres, yet in such case the *real* longitude included within the apparent limits of the chord would be different for the two spheres, as such real longitudes would be in proportion to the difference of the diameters, or rather the bulk, of the two spheres.

Neither would the *appearances* of the oscillation of the longitude of the two spheres be the same, for the reason that on the larger sphere the apparent oscillation would include but its radius, while on the smaller it would include its entire diameter, and such oscillation would thus correspond, in proportion, to the relative bulk of the two spheres, represented by the respective diameters of the two spheres. And such difference in the ratio of the oscillations would also be apparent were the oscillations less than the radius and diameter of the spheres.

As the distance of the sun, by this method of Leverrier's, is measured by the extent of the *apparent* oscillation observed on a longitude of the sun corresponding—by a law of proportion—to the real oscillation of the earth, how could the two *different* appearances, in the case of the two spheres, measure the *same* distance.

In the case of these two spheres, it will be conceded that these *different* appearances of oscillation could not give the *same* results in accordance with Leverrier's method, hence, unless the *real* extent of the apparent longitude were known, the distance of the sun could not be found by this method, and this is neither pretended to be known, nor indeed is it included in the requirements of this method for finding the sun's distance.

For as the *appearance* in the longitude of the sun is, by this method, the real key which unlocks the distance, how can the method be, *per se*, reliable, when such appearances are different in different bodies, in cases where such bodies are at the same distance, for this would be equivalent in want of reliability were the appearances the same, yet the bodies were really at different distances.

Prof. Newcomb, in his "Popular Astronomy," says :—
"Another recondite method has been employed by Lever-
"rier. It is founded on the principle that where the rela-
"tive masses of the sun and earth are known, the distance
"can be found by comparing the distance which a heavy
"body will fall in one second at the surface of the earth
"with the fall of the latter towards the sun in the same
"time."

To which theory, the following criticism may, possibly, be an answer, namely :

How far the earth would fall towards the sun, in a second, or in any other given time, would depend on the quantity of the matter constituting the sun, together with its distance from the earth, because the fall of all bodies is a question of gravitation, and the amount of it, in this case, must depend on the quantity of the matter of the sun compared with that of the earth, and the square of the distance interposed between these two bodies.

As to the mass or weight of the sun, the spectroscope reveals there enormous masses of fiery vapor, and oceans of liquid fire, in fact, these constitute the chief phenomena observed, while the proportions of the weight of these to the whole body is not known, indeed, is not pretended to be known, nor even conjectured.

For this reason, if for no other, the *real* attraction of the sun upon the earth must, at best, be but conjectural, hence, "the relative *masses* of the sun and earth" cannot be known, and upon the real knowledge of these relations must the reliability of this method greatly depend.

As a measure of its attraction, the real distance of the sun can only be known—according to the law of attraction between masses—by squaring the distance intervening between the earth and sun, and how can this be done, when such knowledge is admitted by astronomers to be a problem ?

Knowing neither the real weight of the sun, nor its distance, how far the earth would fall towards the sun in a second could not be known otherwise *than* relatively or proportionately on an assumed basis.

CHAPTER II.

MEASUREMENT OF THE DISTANCE OF THE SUN.

The present writer has found by experiment that by virtue of a law of proportion, observed by him to exist between large and small right angled triangles, under certain conditions, and by means of certain formulas, the hypothenuse and perpendicular of the larger triangle may be found.

These conditions are as follows, namely:

1. The base line of the larger triangle must be known.

2. The three sides of the smaller triangle must also be known.

3. The base line and the hypothenuse of the smaller triangle must coincide with, that is to say, must be parts of the base line and of the hypothenuse of the larger triangle as far as these lines of the smaller triangle extend.

4. The perpendicular of the smaller triangle must be parallel with the perpendicular of the larger triangle.

By observing these essential conditions, and by proceeding in accordance with the following methods, the hypothenuse and perpendicular lines of *any* right angled triangle may be found.

The hypothenuse line of the right angled triangle shown on Fig. 2, may be found by this method by proportion, in accordance with the following formula, namely:

As the base line 1, 2, of the smaller triangle 1, 4, 2, is to the base line 1, 3, of the larger triangle 1, 5, 3, so is the hypothenuse 1, 5, to the hypothenuse 1, 4.

The calculation here is made in twelfths of inches, as follows:

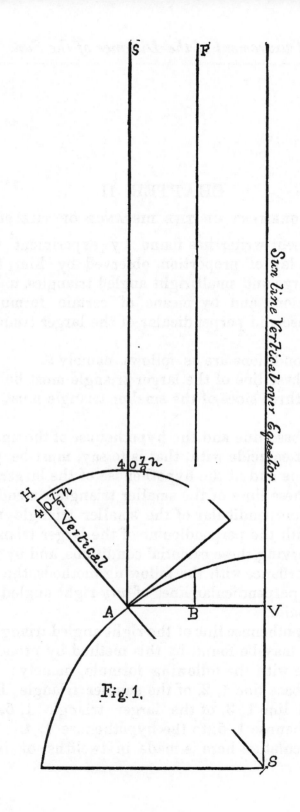

Fig. 1.

Measurement of the hypothenuse line 1, 5, of the larger triangle 1, 5, 3, (Fig. 2,) will show its length to be 128 twelfths of inches.

Base,		Base,		Hyp.		Hyp.
1, 2,		1, 3,		1, 4,		1, 5,

$$12 : 48 :: 32 : ——$$

$$
\begin{array}{r}
48 \\
\hline
256 \\
128 \\
\hline
12)\overline{1536} \\
\hline
128 \text{ twelfths.}
\end{array}
$$

The following is the formula for finding the length of the perpendicular line 3, 5, of the triangle 1, 5, 3, (Fig. 2). The calculation is made in twelfths of inches, namely :

As the hypothenuse line 1, 4, of the smaller right angled triangle 1, 4, 2, is to the hypothenuse line 1, 5, of larger right angled triangle, 1, 5, 3, so is the perpendicular line 3, 5, to the perpendicular line 2, 4.

Measurement of the perpendicular line, namely, the line 3, 5, of the larger triangle 1, 5, 3, (Fig. 2,) will prove the length of such line to be 120 twelfths of inches.

Hyp.		Hyp.		Base,		Base.
1, 4,		1, 5,		2, 4,		1, 2,

$$32 : 128 :: 30 : ——$$

$$
\begin{array}{r}
30 \\
\hline
32)\overline{3840}(120 \\
32 \\
\hline
64 \\
64 \\
\hline
0
\end{array}
$$

These measurements demonstrate the fact that where the conditions before mentioned exist, and the formula is employed, as shown in connection with Fig. 2, that the hypothenuse and perpendicular lines of a larger right angled triangle may be found by means of the short lines of a smaller one as the result of the method by proportion, as just shown.

This fact will also be demonstrated in the same way, namely, by measurements, in connection with the following Figs. 3 and 4.

For example, suppose the lines 1, 3, and 4, 5, on Fig. 3, to be the opposite shore lines of a water course, in miniature, intervening between these lines.

On this diagram, the larger triangle is formed by the lines 1, 4,—3, 4,—1, 3, and the smaller triangle by the lines 1, 2,—9, 1,—2, 9.

The width of this miniature water course is required by by the proportional method, along the dotted line 3, 4.

The line 1, 3, is the shore line, being also the base line of the larger triangle, 4, 1,—3, 1,—4, 3.

The width of this miniature water course may be found as follows, in twelfths of inches, namely :

As the length of the line 1, 2, equal to 18 twelfths of inches, is to the base line of the larger triangle, equal to 84 twelfths of inches, so is to the length of the line 2, 9, equal to 9 twelfths, to the width of the stream along the dotted line 3, 4.

	Base,	Base,	Per.	Per.
	1, 2,	1, 3,	2, 9,	3, 4,

Measurement of the miniature stream along the dotted line 3, 4, shows its width to be equal to 42 twelfths of inches.

$$18 : 84 :: 9 : ——$$

$$\begin{array}{r} 9 \\ \overline{} \\ 18)756(42 \\ 72 \\ \overline{} \\ 36 \\ 36 \\ \overline{} \end{array}$$

But again suppose that each 12th of an inch of the lines employed to find the width of the miniature stream really represented 5½ yards, equal to a rod in distance or length.

On such supposition the real length of the base line of the larger triangle would be 84x5½, equal 462 yards.

The line 1,2, would be 18x5½, equal 99 yards.

The line 2,9, would be 9x5½, equal 49½ yards.

In such case the real width of the stream as found in accordance with this reduced scale would be as follows :

As the width of the miniature stream is in twelfths of
inches, each of these would
be required to be multiplied
by 5½ yards, because the real
length of the base line is estimated in yards. thus, 42 x 5½
equal to 231 yards, would be
the width of the stream.

Base,	Base,	Per.	Per.
1, 2,	1, 3,	2,9,	3,4,

$$99 : 462 :: 49½ : —$$

$$
\begin{array}{r}
49½ \\ \hline
4158 \\
1848 \\
231 \\ \hline
99)22869(231 \text{ yds.} \\
198 \\ \hline
306 \\
297 \\ \hline
99 \\
99 \\ \hline
\end{array}
$$

In the *actual* measurement of a real river, estuary,
sound, lake, etc., the smaller triangle required for such
measurement, say, for instance, such as the smaller triangle
1, 2, 9, Fig. 3. may be as small as may be required for
the measurement of its sides, provided the respective
sides of such triangle *are* of the same denomination as
that adopted for the measurement of the base line of the
larger triangle, that is to say, if the length of the base
line be calculated in inches then the lengths of the lines
of the smaller triangle must be calculated in inches.

If the base line of the larger triangle be calculated in
rods, yards, miles, etc., so must the smaller.

If the width of water be actually measured—or any
horizontal distance sought,—say the width of a river is
required by actual measurement according to the proportional method, proceed as follows :

Say for an example the width of the river were required from 3 to 4 Fig. 3. proceed along a straight stretch of
shore line such as the line 3,1, any convenient number of
yards or rods, say one, two, or 300 yards to 1.

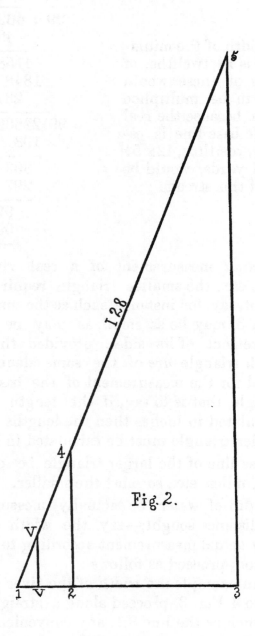

Fig. 2.

At 1 drive a stake vertically at the water's edge. Let a straight edge represented by 1, 9, be pivoted and made to turn horizontally on the top of such stake and parallel to the surface of the water, and near such surface.

Let a person sight along this straight edge—represented by the line 1, 9,—on a tree, bush, rock or other object located at the edge of the water at 4, on the opposite shore. When this is done, draw the line 2,9, perpendicular to the straight shore line. and the width of such river may be found in accordance with the formula shown by the last preceding calculation.

The correctness of this reduced scale method—shown in connection with Fig. 3—for finding greater distances by means of short lines may be proved by reference to Fig. 2.

The perpendicular 2, 4, of the smaller triangle 1, 4, 2, is found in the thirds of inches thus :

As 1, "V, equal to ⅓ of an inch, is to 1, 2, equal to 3 thirds of an inch, so is the line "V, V, equal to 2½ thirds of an inch, to the perpendicular 2, 4.

Calculation by the method by proportion finds the length of the line 4, 2, to be 7½ thirds of inches, equal to 2½ inches, and measurement of the perpendicular line 2, 4, shows such to be its length.

But the length of the perpendicular 3, 5, of the larger triangle 1, 5, 3, is required by the reduced scale method which found the real width of the stream Fig. 3 by knowing and employing the real length of the base line of the larger triangle 1, 5, 3. Fig. 2.

Measurement of this base line will show its length to be 4 inches, namely 4 times the length of the base line of the smaller triangle. In such case, as in the measurement of the width of the stream, the reduced scale would be required to be increased proportionately in order to find the real length of the perpendicular line 3, 5. As the base line of the larger triangle is 4 times that of the smaller, as measurement of it will show, hence according to the law of proportion 1,"V, would represent 4 thirds of inches, instead of one third as it does by the reduced scale.

The line 1, 2, would represent 12 thirds instead of 3 thirds, and the line "V,V, would represent 10 thirds instead of 2½.

The calculation in thirds of inches, to find the length of 3, 5, will be as follows :

As 4 is to 12 so is 10 to 3, 5.

$$4 : 12 :: 10$$
$$\underline{10}$$
$$4)\overline{120}$$
$$\underline{3)\overline{30} \text{ thirds.}}$$
$$\overline{10} \text{ inches.}$$

Measurement will show that the length of 3, 5, the perpendicular of the larger triangle (Fig. 2,) is 10 inches.

This measurement corroborates the correctness of the measurement—by the same method—of the width of a supposed river in connection with Fig. 3, and proves that where the real length of the base line of a larger triangle is known, and employed in the measurement, by following the method before described and adopted in connection with Figs. 2 and 3, that lengths of long lines may be found by means of short ones.

In order to demonstrate mathematically the correctness of this proportional method when applied on a reduced scale to find greater lengths of lines and greater distances, suppose a column to be 80 feet high, and that a base line be drawn from the central line of the column, at its base, and at a right angle to the perpendicular line of the column, a distance of 60 feet.

From the terminus of this base line furthest from the column, it is required that a wire be extended to the central line of the column at its top.

What is required to be known is the length of wire necessary in the case supposed.

In order to ascertain this, draw such column in miniature, as shown on Fig. 4. The dotted line 1, 2, is 4 inches long. The base line, 3, 2, is 3 inches long.

As the column is 80 feet high by supposition, and the miniature column, shown on Fig. 4, is 4 inches high, each inch of the reduced scale on which the miniature column is drawn—as also its base line—represents 240 inches, equal to 20 feet, as 240 x 4 equal 960 inches, and 960 divided by 12 equals 80 feet, and so for the base; as 240 x 3 equals 720, and 720 divided by 12 equals 60 feet, the length of the base line of the real column.

In finding the height of the miniature column, on Fig. 4, the calculation below is made in twelfths of inches, and is in accordance with the following formula, namely:

As 6 twelfths, namely, the length of 3, 7, of the base line of the smaller triangle, are to 36 twelfths, namely, to the length of the base line of the larger triangle, so are 10 twelfths, namely, the hypothenuse line of the smaller triangle inscribed within the larger, to the hypothenuse line of the larger, thus:

The hypothenuse 3, 1, of the larger triangle, 3, 1, 2, on Fig. 4, is then 60 twelfths of inches long, equal to 5 inches, and measurement will prove such to be its length.

$$
\begin{array}{cccc}
\text{Base,} & \text{Base,} & \text{Hyp.} & \text{Hyp.} \\
3, 7, & 3, 2, & 3, 1, & 3, ``1 \\
\hline
6 & : \ \overline{36} & :: \ \overline{10} \\
& 10 \\
& \overline{} \\
& 6)360 \\
& \overline{} \\
& 12)60 \text{ twelfths.} \\
& \overline{} \\
& 5 \text{ inches.}
\end{array}
$$

But in accordance with the reduced scale on which Fig. 4 is drawn, each linear inch on this diagram represents 20 feet, equal to 240 inches.

Then as five inches, namely, the measured length of the hypothenuse line 3, 1, on Fig. 4, multiplied by 20 feet is

3

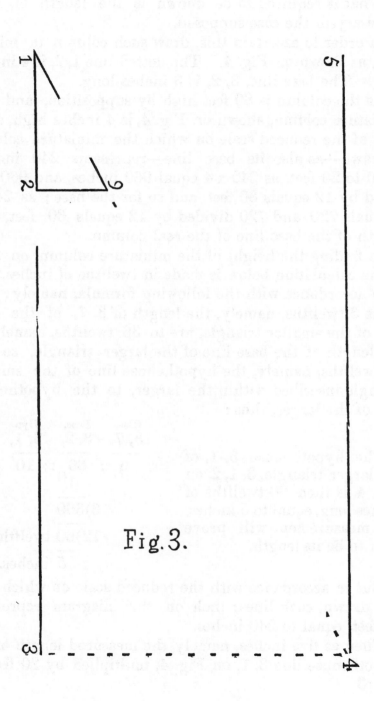

Fig. 3.

equal to 100 feet, 100 feet will be the length of wire required from the terminus of the 60 feet base line to the apex of the 80 feet column.

Let the correctness of this conclusion be tested by the 47th Proposition of the first book of Euclid.

This Proposition—demonstrated to be correct—is as follows, namely: "The square of the hypothenuse of a "right angled triangle is *equal* to the sum of the square of "the two other sides."

The triangle forming Fig. 4—which represents in miniature the direction of the column, its base line and hypothenuse is right angled; that it is so, may be as well—and more quickly—found by applying an ordinary square to the lines on Fig. 4, as by a technical demonstration.

The sides 1, 2,—3, 2, that is to say, the height of the column and the length of the base line, are known, as stated.

As the column is 80 feet high, the square of this side of the right angled triangle is 80 x 80 equals 6400.

The square of the base line is 60 x 60 equals 3600.

The sum of the square of these two sides is then 6400 and 3600, equal to 10,000.

Then the square of the hypothenuse of this right angled triangle, namely, that which the line 3, 1, represents on Fig. 4, must be 10,000, because, as just shown, 10,000 is the sum of the square of the other two sides of the right angled triangle; were the square of its hypothenuse more or less than 10,000, it would be more or less than *equal* to the sum of the square of the other two sides.

When by the terms of the 47th Proposition the square of the hypothenuse is *equal* to the sum of the square of the other two sides of a right angled triangle.

Then the square of the hypothenuse of the triangle, here supposed, is 10,000, as this is *equal* to the sum of the other two sides, which is in accordance with the requirements of the Proposition.

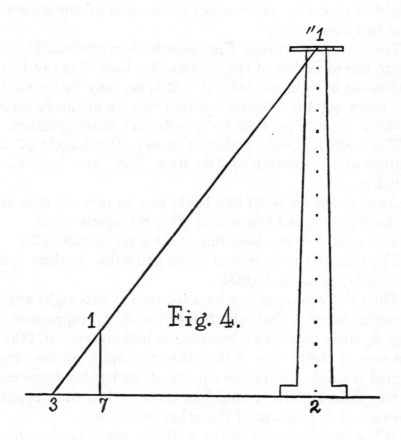

Fig. 4.

10,000 results from the square of 100, as 100 x 100 equals 10,000.

By extracting the square root of 10,000 it will be found to be 100, and for the reason just given, namely, that 100 x 100 equals 10,000.

Measurement of the lines of the right angled triangle, on Fig. 4, will prove by *analogy* the correctness of the foregoing deduction, namely, the base line 3, 2, is 3 inches, the square of this is 9. The line 1, 2, is 4 inches long, the square of 4 is 16. The sum of the square of these two sides of this right angled triangle, namely, 9 and 16, is 25. Then the square of the hypothenuse 3, 1, must be— in accordance with the terms of the 47th Proposition—25. The square root of 25 is 5, as 5 x 5 equals 25 ; then the hypothenuse line 3, 1, should be five inches long, and measurement shows it to be so.

As this 47th Proposition finds the length of wire required—in the case supposed—to be the *same* as that found by the proportional method here employed for the same purpose, the following conclusions are deduced, namely :

1. That the hypothenuse of a right angled triangle may be found by the proportional method, in all cases where the base of such triangle is known, and the same formula is employed for that purpose as that by which the hypothenuse line was found of the right angled triangle in Fig. 4.

2. That when this method is employed on a reduced scale, by means of short lines, the lengths of long ones may be found, as it has been shown shown in connection with Fig. 4, that by lines respectively 3 and 4 inches long was found the length of a hypothenuse line 100 feet long.

3. That the *different* lengths of lines—representing on a reduced scale the longer lines sought by the method— does not affect the reliability of the method, seeing as we do, that the short lines representing the longer lines, or greater distances sought, differ in lengths on Figs. 2, 3, 4,

also on Fig. 5—as also do the lines so found differ in lengths—yet the distances or lengths sought by this method are correctly found as proved by measurements of them as shown in connection with all the diagrams just numbered.

By employing this method on a reduced scale in the same manner as that which found the lengths and distances sought in connection with these diagrams, the present writer has measured the distance of the sun as is hereinafter shown.

In order to do this with any reliability in the result found, it is necessary to show that the method employed is substantially the same as that which found the lines sought on the different preceding diagrams.

The first prerequisite to the measurement of the sun's distance, by this method, is the construction of the larger right angled triangle, and preliminary to this is the necessity of finding the proper angle at the point of junction of the hypothenuse with the base line of such larger right angled triangle.

It will be observed, on Fig. 1,—and measurement will show its correctness—that to an observer at A, 40° N., the line A S is 40° S. of the vertical line A H, as measurement will show that the line A S cuts the celestial arc H M 40° S. of H. H being the zenith over A. *See Fig.* 1.

Were the line A H placed vertically on the 21st of March or September, the line A S should on such day point on the centre of the sun, because A being 40° N. of the equator and the sun on such day being 40° S. of A were A S pointed 40° S. of A it should point on the sun.

In order to be practically certain of the correctness of this theoretic conclusion, and more especially to test the reliability of the diagram, Fig. 1,—which diagram is intended to represent in miniature a quadrant of the earth's circle—the following experiment was made by the writer, with a duplicate of Fig. 1, on the 21st of September, at 40° N.*

On that day the writer placed a low table on the ground, its top being horizontal, so found by a spirit level.

A North and South line was drawn on the top of this table by a compass. A line, at right angles to this, was drawn, representing the East and West line.

A duplicate of Fig. 1 was smoothly attached properly to the West side of this table, coincident with a North and South line, the arrow S on Fig. 1 pointing in a southerly direction.

The line C A H on Fig. 1 was placed vertically, such line coinciding in direction with the plumb line.

A straight edge was securely fastened to Fig. 1, coincident in direction with the line A S and a short distance in extension of such line.

A few minutes before the sun was on the meridian line, the writer sat on the ground, and sighted from A, along the straight-edge just mentioned, through an apparently properly smoked glass of good material, and but a very short time before meridian time the sun commenced to cross the line of sight.

The sun appeared to cut quite clearly across the line of vision directed along the straight-edge, and *nearly* coincident with the horizontal diameter of the sun.

It is claimed theoretically by the direction the line A S is drawn, namely, 40° S. on Fig. 1,—and practically as a result of this experiment—that the line A S, Fig. 1, is drawn at the proper *angle* to point centrally on the sun when the sun is vertically on the equator, and the observation is made from the point A 40° N. on the 21st of September; although such line pointed a little *below* the centre of the sun, but "owing to refraction the true alti-"tude is always *less* than the observed."

---

* In order to avoid repetition of fractions, the distance from the equator where the experiment was made is stated to have been at 40° N., whereas the exact distance is 40° 7–9, and Fig. 1, was drawn from 40° 7–9.

According to the law governing the refraction of heavenly bodies, namely, nothing on the zenith and greatest on the horizon, at the altitude the sun then was, namely, 40° above the southern horizon, its real altitude below the apparent would have been 54″, in which case the line of vision along the straight-edge would have been coincident with a horizontal drawn through the sun's centre.

From this experiment the deduction is made that the hypothenuse line A S, Fig. 1—which is intended for the hypothenuse of the larger triangle hereinafter explained—is drawn at the proper angle to reach the sun's centre, were such line indefinitely continued in that direction.

The larger right angled triangle, necessary to the measurement of the distance of the sun by the method by proportion, is formed as follows :

1. By the hypothenuse line A S, Fig. 1, extended by and coincident with the line of vision to the centre of the sun from 40° N. at the times of the equinoxes.

2. The perpendicular of such triangle is the vertical line from the sun's centre to the earth's centre through the equator at the time of the equinox.

3. The base line of such triangle is the measured distance of a line drawn from such vertical line, and at a right angle to it, to the point A 40° N.

The fact that no *real* lines, forming such larger triangle, extend from A to the sun, from the sun vertically through the equator, nor from the vertical line (drawn to the centre of the earth) to the point A, 40° N., cannot detract from the mental contemplation of such triangle ; for were the lines under consideration capable of being constructed, it would be necessary to construct them along the line of vision. This line of vision enables us to mentally construct such lines as perfectly to the conception as though they were actual material lines.

To the imagination, that invisible eye of the mind, a

triangle may appear as clear and distinct, as though drawn by the hand, and observed by aid of the visual organ.

If a line be drawn by the imagination from 40° N. to the sun's centre at the period of the equinoxes, and if, at the same time, a vertical line be drawn, by the same intangible and impalpable artist, from the centre of the sun to the centre of the earth, we have presented to our mental view the respective termini of the hypothenuse and perpendicula of two lines, the apex of both lines being at the sun's centre, and the other terminus of each line located in different places on the earth.

If, then, a line be supposed to be drawn at a right angle to such vertical line to the 40th parallel North, say, for example, to the point A, Fig. 1, we have to the mental conception a large right angled triangle, as clear and distinct to the imagination as it could be to the visual organ were it constructed of material lines, and the vision strong and far-reaching enough to take its measure at a glance.   For we apprehend that it will be conceded that the mental capacity to clearly conceive the existence of a right angled—or any other form of—triangle, and the physical capacity to draw the lines constituting it, are so wholly distinct, that the former capacity may exist independent of the latter.

If any support to this deduction were necessary it may be found in the fact that it is possible to clearly conceive of celestial and terrestrial equators, meridians, parallels and tropics—and these do not exist but as mental conceptions—yet these mere ideal creations are found sufficient for the purposes required of them in astronomy, geography, and navigation, as though they had a material existence, while in point of fact it would be as impossible to draw *material* lines in the instances just cited as it would be to draw material lines thus forming the larger right angled triangle now under consideration, hence, as in the instances just mentioned, such triangle may exist in the contemplation of the mind as though it were actually constructed, provided such triangle would exist were it actually capable of real construction.

Such larger right angled triangle may be comprehended by the following tracing of the lines forming it, namely :

1. Its hypothenuse extends from 40° N. to the sun's centre.

2. The vertical line on the 21st of September extends from the centre of the sun to the centre of the earth, through the equator.

These two lines cannot be parallel because, "No two "bodies can approach the centre of a sphere from different "points on parallel lines," hence, as the second named line is vertical, the first named line must be at an angle to the second, because drawn from different points.

3. The measured length of the base line extends from the vertical line, and at a right angle to it, to the point A Fig. 1, 40° N.

These lines form the larger triangle necessary to the measurement of the distance of the sun from the 40th parallel N., in accordance with the method by proportion and upon a reduced scale.

According to the proportional method, in order to find the hypothenuse or perpendicular of any right angled triangle, it is a *sine qua non* that the real length of the base line of any such triangle must be known. This fact is shown by the method and the measurements employed in connection with Figs. 2, 3, 4, and the following Fig. 5.

Hence in order to measure the distance of the sun by this method it is necessary to know the length of the base line of the larger triangle represented in direction and termini on Fig. 1, by the line A V.

There are two methods by which the length of this base line may be found.

1. By finding by calculation the proportions existing between an arc and its chord.

Here the arc would extend, along the curve of the earth's surface, from 40° N. to 40° S. The measure of this curve is known. The one-half of the chord of this arc would be the distance, from A, Fig. 1, 40° N. on a straight line and at a right angle to the perpendicular equatorial vertical line.

2. This distance may be found by drawing a circle and finding by measurement the *proportion* of the segment (extending from 40° N. to the vertical equatorial line) to the straight line extending from the vertical equatorial line, and at a right angle to it, to the point A, 40° N. of the equator.

The writer finds the length of this base line shown in direction by A V, Fig. 1, to be 2562½ miles, without taking into account the slight change in these figures which the oblateness of the earth would cause. 2562½ miles is, in any event, at least a close approximation to the real length of the base line of the larger triangle from 40° N.

In accordance with the method by proportion it is necessary to construct a smaller triangle within the larger, as shown on the different diagrams. Such smaller triangle, although it *was* really constructed, is not shown on Fig. 1, for the reason that the approach of the hypothenuse line A S towards B F on Fig. 1 is so gradual that the point of intersection—forming the apex of the smaller triangle —of these lines is so distant that the smaller triangle cannot be conveniently shown within the limits of a publication such as this.

However, this smaller triangle was constucted within the limits of the larger before mentioned, and in accordance with the method by proportion, as follows, namely :

Fig. 1 was laid horizontally on a smooth board and a pin driven into it at the . dot at B. A fine silk thread was then fastened to the pin and extended over the line B F and coincident in direction with it more than 500 inches.

The line A S was extended in the same manner, and by means of a similar silk thread, more than 500 inches from A.

By this means, the silk thread coincident with and in extension of the line A S, gradually approached and finally crossed the silk thread which extended the line B F, as just stated.

The smaller triangle inscribed within the larger, before mentioned, was thus formed. The hypothenuse line of the smaller triangle being A S extending from A, Fig. 1, to the point where it intersects the line B F, as just stated, while its direction is coincident with the hypothenuse line of the larger triangle which latter is the line of vision from 40° to the sun, as shown by the experiment before detailed.

The perpendicular of such smaller triangle being the line F B, Fig. 1, parallel to the vertical equatorial line V S, its base line being A V.

A better, and a correct, idea of this smaller triangle, on Fig. 1, may be had by reference to Fig. 5, which, except in size, is an exact duplicate of Fig. 1, provided that the imagination supplies the silk threads mentioned in connection with Fig. 1, and not shown on it.

By comparing the method employed for finding the hypothenuse line A S, Fig. 5, with the method which found similar lines on Figs. 2, 3, 4, it will be found that the method in each case is the *same.*

When it is considered that the following formula of the method, applied to the finding of the hypothenuse line, Fig. 5, namely : "As A B is to A V so is A F to A S," does find such a line as proved by measurement of it, it were to be expected that the same formula applied to find the length of the hypothenuse line A S, Fig. 1, would be competent to find the length of such line, and consequent distance of the sun from 40° N.

Indeed, except in the greatly different lengths of the hypothenuse lines of the smaller triangles of Figs. 1 and 5, and in the greatly different lengths sought by them there is no difference either in the forms of the two diagrams, or in the method by which such similar lines in each have been sought.

It will be conceded that were the sun really at S, Fig. 5 —which might be conceived of by a *very* strong effort of a very brilliant imagination on dwarfing its proportions— the distance of the sun might be found along the hypothenuse line A S ; measurement of this line on Fig. 5 shows this.

Measurements of the *different* length of the hypothenuse lines on the preceding diagrams show that the *length* of such lines does not affect the reliability of the method by which they were found.

Would the mere *length* of the hypothenuse line of a right angled triangle—were it millions of miles long— prevent the geometer from finding its length by application of the 47th Proposition of the first book of Euclid, were the other two sides of the triangle known ?

Yet the reliability of the proportional method to find such line— in triangles of different sizes—is demonstrated, on Fig. 4, by this 47th Proposition.

In point of fact, the differences in the relative sizes of the larger and smaller triangles makes no difference in the reliability of the method for finding the hypothenuse—or perpendicular—lines of the larger triangle.

Measurements of these lines on the different larger triangles on Figs. 2, 3, 4 and 5 prove this.

For instance, on Fig. 2, the lines of the larger triangle enclose an area 130 times that of the smaller. On the reduced scale on which Fig. 4 is drawn, the ratio between the height of the column whose height is sought and that of the miniature column is as 240 to 1.

These facts induce the conclusion that the relative size of the smaller to the larger triangle on Fig. 1 will make no difference in finding the sun's distance by the method.

If it be conceded that were the lines A S and B F, Fig. 1, continued to the point of their intersection, as stated, that the smaller triangle would be in all respects—except in size—exactly that shown of Fig. 5—and drawing these unfinished lines will show such to be the case—a comparison of Fig. 5 and its method employed for finding the hypothenuse of its larger triangle with Figs. 2, 3, 4, will show an identity in *all* things required by the proportional method for finding the length of the hypothenuse of larger triangles.

Hence, the length of the hypothenuse line A S on Fig. 1—and consequent distance of the sun from 40° N.—may be found by the *same* method.

That the method employed for finding the length of the hypothenuse line A S, Fig. 1,—extending from 40° N. to the sun—*is* the same as that employed on Fig. 5 for a similar purpose, by a similar line, and is indeed the *same* on all the diagrams, may be shown, say for example, by comparing Fig. 2 with Fig. 1.

On Fig. 2 the length of the line 1, 2, being part of the base line 1, 3, of the larger triangle, is to 1, 3, as 1, 4, the hypothenuse of the smaller triangle, is to 1, 5, the hypothenuse of the larger triangle.

On Fig. 1 when the length of the hypothenuse A S is sought it will be found by comparing Fig. 1 with Fig. 2 —or with Fig. 5—supposing that the lines A S and B F to be extended to the point of their intersection as stated, that such lines correspond on these diagrams, and the method employed to find the length of the hypothenuse line A S will be the same as that employed for a similar purpose on Figs. 2, 5, etc., namely :

The real length of the line A B, being part of the base A V of the larger triangle—whose hypothenuse extends from 40° N. to the sun—is to the *real* length of such base line A V as A F (for this line see Fig. 5,)—being the hypothenuse of the smaller triangle—is to A S, namely, from A, 40° to the sun.

In attempting to find the distance of the sun, a similar reduced scale is employed in connection with Fig. 1, as that employed in connection with Fig. 3 to find the real width of a river, and Fig. 4 to find the real length of a required wire, which method found the real distances and length sought by attaching to the measurement of the scale the real distance and lengths which the reduced scale represented; this may be found to be the fact by reference to Figs. 3, 4.

On Fig. 1 each linear inch of *all* lines required by the method represents 1000 miles in distance.

As the *real* length of the base line, in all cases constitutes a necessary factor in order to find the distance sought by the proportional method, on so small a diagram as Fig. 1 an inch to 1000 miles is necessary in order to properly represent the 2562½ miles which is real distance, and consequently the real length of the base line of the larger triangle. Such base being taken from 40° N. to the vertical line over the equator and at a right angle to such vertical at the time of the equinoxes.

The length of the hypothenuse line A S, Fig. 1, of the larger triangle may be found, say, for example, in sixteenths of inches. Then reduce such sixteenths to inches. Then attach to these inches their values—as representing distances—in accordance with the requirements of the reduced scale on which Fig. 1 is drawn.

Or the length of this line, A S, may be found the same'—as was done in connection with Fig. 4—by attaching the *real* values to the lengths of the respective necessary lines, according to the scale, and then finding the length of the hypothenuse line A S in miles.

Each of these formulas has been adopted in the calculations, as may be seen as follows :

First, in sixteenths of inches, namely, Fig. 1.

.As **A B** is to **A V** so is **A F** to **A S.**

$$20.\tfrac{50}{100} \; : \; 41.00 \; :: \; 8000 \; : \; ——$$

$$8000$$

$$20.\tfrac{50}{100})32800000(16000$$
$$2050$$

$$12300$$
$$12300 \qquad\qquad 16)16000$$

$$000 \qquad\qquad 1000$$

According to the above calculation the full length of the line A S—accrding to the reduced scale on which Fig. 1 is drawn—is 16,000 sixteenths of inches, equal to 1000 inches, having a value—according to the scale—of 1000 miles to each inch, hence in accordance to the requirements of such reduced scale—being the same as that found competent to find long hypothenuse lines by means of short ones, as proved correct in connection with Figs. 3 and 4—this is equivalent to 1000 inches multiplied by 1000 miles, which is equal to 1,000,000 miles, such being the real distance of the sun from 40° N. taken at the time of the equinoxes.

The distance of the sun, from 40° N., estimated in miles, is as follows :

As **A B** is to **A V**   so is   **A F**   to   **A S.**

$$1281\tfrac{25}{100} \text{ miles} : 2562\tfrac{50}{100} \text{ miles} :: 500,000 \text{ miles}* : \; ——$$

$$500000$$

$$1281.25)128125000000(1,000,000 \text{ miles.}$$
$$128125$$

$$000000$$

---

* This 500,000 miles represents, in accordance with the scale, the 500 inches, —namely, 1000 miles to the inch—measured from A, Fig. 1, by the silk threads before mentioned, coincident with A S to the point of intersection of the lines A S and B F.

Consequently, the distance of the sun from the surface of the earth, at 40° N., is one million miles.

If the silk threads, extending the two lines,—namely, A S and B F, Fig. 1,—actually extend these lines to the point of their intersection, they will be found to intersect 500 inches from A along the line A S.

If this be done—of even supposed to be done—it will be found that there will be no difference in the structure of the triangles Figs. 5 and 1, and no difference between them except that the triangles of the latter are larger than the former.

Inspection of the two diagrams will show this to be the case; and by observing the diagrams it will be found that the calculations made for finding the lengths of the hypothenuse lines A F S on both these diagrams are by the same method as that employed successfully to find the hypothenuse lines on Figs. 3 and 4, which has been demonstrated, in connection with Fig. 4, to be *per se* correct, and also proved by measurement to be capable of finding the lengths of longer hypothenuse lines not shown on these diagrams, by means of the short ones shown on them, just as the same method has been employed by means of Fig. 1 to find the long hypothenuse line extending from the point A to the sun.

As in the other examples—as also in connection with

Fig. 1—the length of the hypothenuse line A S, Fig. 5, may be found as follows, namely,

As A B is to A V so is A F to A S; calculated in eighths of inches.

$$\overset{\text{A B}}{4.50} \; : \; \overset{\text{A V}}{9} \; :: \; \overset{\text{A F}}{35} \; : \; \overset{\text{A S}}{\underline{\quad\quad}}$$

$$35$$

$$4.50)315.00(70 \text{ eighths.}$$
$$315.0$$
$$\overline{\qquad\qquad}$$
$$0$$

Measurement of the line A S, Fig. 5, shows its length to be 70 eighths of inches.

This identity of form, method and form of calculation between Figs. 1 and 5 induces the necessary conclusion, that as the hypothenuse line of Fig. 5 is found by measurement to be of the same length as that found by the method, that were the hypothenuse line to the sun, Fig. 1, susceptible of measurement, its length so found would also prove the correctness of the method; hence, that the method employed in connection with Fig. 1, finds the real distance of the sun.

Should the distance of the sun be required along the vertical line V S, Fig. 1, proceed in accordance with the formula employed to find the perpendicular line 3, 5, on Fig. 2.

The distance of the sun may be found by the method shown in connection with the different diagrams from any latitude, as follows:

Find some mountain peak—or other elevation high above the earth's surface—far enough North or South of the equator to form some considerable angle to the line perpendicular to the equator at the time the sun is vertical over it.

Let the observer be on the meridian line far enough from such peak that when sighting over its top when the sun is at its meridian height at the time of the equinoxes —say on the 21st of September—the line of sight of such observer will be—from such point of observation coincinent in direction with the degree of longitude on which such observation is made—pointed centrally on the sun and just over the top of such mountain peak.

Let the place of the sun's measurement be on the *slope* of a hill having a *Southern** exposure.

If none such can be found at a place otherwise suitable for the measurement of the sun's distance, such slope as hereinafter explained might be temporarily constructed at but little cost.

In order to find the direction of the base lines of the triangles at the time the sun is vertical over the equator, at a right angle to such vertical line :

1. Have a thin metallic semi-circle constructed, say, for example, 12, 24 or 36 inches in diameter, having degrees marked on a flat side of the semi-circle and on the edge of its circumference ; and having a movable hand or pointer pivoted midway the diameter of such semi-circle, so that such hand may be moved over the degrees on it, as the minute hand of a clock moves over its dial.

Let the edge of the diameter of this semi-circle be placed on the earth's surface so that this edge will be horizontal ; such plate standing on its diameter edge being vertical on a North and South line, and its pointer, or movable hand, being on a horizontal line and close to the surface of the earth.

Suppose this movable hand, from its horizontal position, its point pointing due South, to be depressed or moved downward towards the earth, round the semi-circle the *same* number of degrees marked on it, as is the place of measurement of the sun's distance from the equator ; say, for example, the place of measurement is 45° from the equator, then move the hand on the face of the semi-circle

---
* If the measurement be made in North Latitude.

downwards towards the earth, from a horizontal position until it points to 45° on the face of the semi-circular plate, and in this latter position this hand of the semi-circle will point at a right angle to the vertical line from the equator to the centre of the sun at the time of the equinoxes.

The *direction* of this hand will then be the direction of the base line of the larger *and* smaller triangles necessary to the measurement of the sun's distance at the time the sun is vertical over the equator.

The length of the base line may then be found by calculation, or by the proportional method as hereinbefore explained.

In order to show the necessity and the efficiency of this pointer hand, movable over the degrees of latitude marked on the semi-circular plate, and of the slope of ground having the Southern exposure, suppose—merely for illustration—that the sun were vertical over the place of measurement of the sun's distance, say over 45° N., and the pointer hand on the semi-circular plate were at a right angle to the sun's vertical line, namely, horizontal.

Then suppose the sun were instantly moved around the circle of the heavens, southward on its meridian line to the point of the heavens at which it would be vertical over the equator of the earth.

As all circles have the same number of degrees, were the pointer hand of the semi-circular plate depressed towards the earth, and thus moved over the face of the plate the *same* number of degrees as the sun had moved, in the case supposed, namely, 45°, it is clear that such pointer hand would be in the direction of a right angle to the sun's vertical line at the equator, as it was at such angle when the sun was at 45° N., in the case supposed.

A little reflection would show that such would be the case at *any* latitude, if such pointer hand be depressed from the horizontal line towards the earth the *same* number of degrees on the semi-circle that the place of measurement of the distance of the sun is from the vertical line of the sun over the equator.

The slope of the ground, having a southern exposure, is necessary to the proper construction of the smaller triangle, in order that its base line coincides in direction with the depressed pointer hand.

For while it would be more easy and convenient to construct the base line of the smaller triangle on the horizontal line, yet it will be perceived were this done, such base line would not coincide with the base line of the larger triangle, as the latter base line would be coincident in direction with the depressed pointer hand, as such would be the direction of the base line of the larger triangle, in order that it be at a right angle to the vertical line of the sun over the equator.

Were the base line of the smaller triangle drawn horizontally, and the base line of the larger drawn at an angle of 45°, as in the case just supposed, it will be perceived that there would be a difference in the directions of the base line of the larger and smaller triangles of 45°, when in accordance with the method by proportion it is necessary that the base lines of both these triangles should be *coincident* in direction, hence the necessity of the slope of ground mentioned,—or of its equivalent—namely, a slope of 45° to the horizon, were the measurement of the sun's distance taken from the 45th parallel N.

The hypothenuse of the larger triangle is the line of sight, directed, from the point of measurement of the sun's distance, just over the top of the mountain peak to the centre of the sun, when the sun is at its meridian height at the time of the equinoxes.

The base line is the distance from such point to the vertical line of the sun over the equator and at a right angle to such line.

The perpendicular is such vertical line extending from

the point of its intersection with the base line to the centre of the sun.

The smaller triangle within this larger is found as follows:

From the pivot of the pointer hand, when the pointer hand is directed 45° below the horizon, as before stated, let a pole be extended, coincident in direction with such pointer hand, namely, in the direction of the equator, and from the further extremity of such pole let the line of sight be directed so that it will be just over the top of the mountain peak, while such line of vision will be at the same time perpendicular to the length of the pole.

The intention of this is *thus* to form the perpendicular line of the smaller triangle.

In some instances, owing to the height of the peak, the distance of the point of observation from it, or the length of the slope southward arising from these causes, a surveyor's chain may be necessary to measure the distance required for the base line, coincident in direction with the depressed pointer hand, in order that the perpendicular line at a right angle to such pole or chain, may be formed.

In this manner of proceeding, the base line of the smaller triangle, and the line perpendicular thereto may be formed.

The hypothenuse of such smaller triangle consists of the line of sight from the point of observation, namely, from the centre of the semi-circular plate at the earth's surface, to the top of the mountain peak, when the sun is at meridian as before stated.

With the foregoing prerequisites ascertained, the length of the hypothenuse may be found by means of the formula shown on Fig. 2, which was employed to find the length of the hypothenuse line of the smaller triangle inscribed within the larger as shown on that diagram.

This formula is as follows, namely:

As ″V 1 is to 1, 2, so is 1, V to 1, 4.

When the hypothenuse line of the smaller triangle under consideration is *thus* found, namely, when such line if extended would reach the centre of the sun, which it would do at the time of the equinoxes, then proceed to the measurement of the distance of the sun.

This may be done in accordance with the formula employed in connection with Fig. 2, to find the length of the hypothenuse line 1, 5, of the larger triangle shown on that diagram. The following formula employed to find the hypothenuse line extending from, say 45°, to the centre of the sun, on comparison, will be found to be the same as that which found the hypothenuse of the larger triangle on Fig. 2, namely :

As the length of that part of the base line of the larger triangle— found by the pole or the surveyor's chain, as the case may be—is to the real length of the base line of the larger triangle (which extends from the point of measurement to the vertical equatorial line) so is the length of the hypothenuse of the smaller triangle (extending from the point of measurement at 45° to the top of the mountain peak) to the distance from the surface of the earth at 45° N. to the centre of the sun.

Great care is necessary, in order to seoure accuracy, in the employment of the proportional method, and on a reduced scale to find the distance of the sun, for the reason that a very small inaccuracy in the line of sight, or in the construction of the smaller triangle, may result in a very considerable error as to the true distance of the sun.

For these reasons there may possibly be—though not probable—error in determining the distance of the sun, in accordance with the method as detailed, but as great care was used, the distance found by it, namely, 1,000,000 miles, may be considered as, at least, a close approximation to the real distance.

But, accuracy, resulting in certainty in constructing the lines of the smaller triangle, and correct observation of

the sun are all that appear necessary to find the distance
of the sun over a suitable mountain peak, when such distance is sought by the proportional method on a reduced
scale.

This conclusion has been reached for the reason that it
has been demonstrated by measurement, also by a mathematical Proposition in connection with Figs. 4, 3, that such
method will find longer distances, greater heights, and
widths, than those of the lines, representing these, and
shown on the diagrams.

These methods may also be employed to find any length
of perpendicular or hypothenuse line which may be found
by the 47th Proposition of the 1st Book of Euclid.

Indeed, it may be said to be an improvement on that
Proposition, for the reason that—by aid of the smaller
triangle, which is easily constructed—all that is required
to find the hypothenuse and perpendicular of *any* right
angled triangle is to know the length of its base line; thus
it will be seen that *two* sides of such triangle may be
found on knowing the length of *one* side by this method
while by the 47th Proposition the knowledge of the
lengths of *two* sides of such triangle is necessary in order
to find *one* side.

While the disproportion between the scale on which
Fig. 1 is drawn, and the value attached to its lines, namely, 1000 miles to the inch, is very great, yet experiment
shows that this makes no difference in the correctness of
the result, so long as the proportional method is observed.

This conclusion may, *pro tanto*, be shown to be correct
by Fig. 4, where the scale is reduced to the $\frac{1}{240}$ part of the
height of the column, whose height is found by it, as has
been before mentioned.

This result induces the conclusion that whatever may
be the extent of the difference between the reduced scale
and the length of the lines represented by it—however

great such difference may be—will not lessen the reliability of the lengths or distances found by it.

· Hence, in attempting to find very great heights or distances by the proportional method and on a reduced scale, the smaller triangle may be as small as is within the limits of practical measurement without regard to the comparative sizes of the larger and smaller triangles.

· As the distance of the sun, according to the conclusion of astronomers, is the unit of measurement by which the distances of the planets must be determined, if this rule be observed, and be applicable, the respective distances of the planets would be required to be reduced, to be in harmony with the reduced distance—and mass—of the sun, as found by the measurement of that distance herein made.

The measurements of the lines drawn by which the lengths of longer lines are sought by the proportional method, and then attaching to such shorter lines the values adapted to the reduced scale on which such shorter lines are drawn—as shown by the different diagrams—*demonstrate*, by this infallible test, namely, by measurement, the abstract correctness of this method for finding long distances by means of short lines.

· On a comparative examination of this method—and its application—by which the distance of the sun was sought from 40° N., with that by which the hypothenuse lines on preceding diagrams were found, the method and its application will be found to be identical.

If, upon such comparative examination, such method and its application be proved to be identical, then it must necessarily follow that the distance of the sun from the earth has been thus *demonstrated*.

If the method so tested be found identical, as stated, and the distance of the sun be not found by it to be as herein shown, then the error must be either :

1. In want of mathematical accuracy in drawing the necessary angular lines on Fig. 1, or,

2. In the vertical line on it, not having been *so* drawn, or,

3. In want of parallelism of the line B F with such vertical line, or,

4. In want of parallelism of the straight edge with itself, or,

5. In want of perfection of the smoked glass through which the observation of the sun was made, as before detailed, or,

6. In want of some requirement, or in some particular, omitted or absent, when the experiment was made by the writer on the 21st of September, as stated.

If any of these causes of error existed—and the writer is of the opinion that none such did exist—they would not affect the reliability of the method *per se*, for peradventure with a perfectly constructed (Fig. 1,) diagram ; perfect instrumentalities, and accurate observation of the sun at the time of the equinoxes, error from the causes enumerated—if any such existed—may be eliminated, and probably, the *exact* distance of the sun may be found.

Or, it is possible for the method to be carefully and accurately applied to the measurement of the distance of the sun, say at 40°, 45°, 50°, or 60° of latitude over some mountain peak suitably situate, and the topography of the surroundings such as would be necessary—or over some other high elevation—as before herein detailed, and thus the exact distance of the sun may be found by the method by proportion and on a reduced scale.

CHAPTER III.

THE DIAMETER, LIGHT, HEAT, AND ATTRACTION OF THE SUN AND THE SOURCE OF ITS HEAT.

As shown in Chapter II, the distance of the sun from the earth is one million miles.

A pertinent question which presents itself is, "What is its diameter, based on such distance ?"

A sun distant one million miles should have a diameter *proportionate* to that of a sun 93 millions of miles distant having a diameter of 882,000 miles, in order that the light and heat emanating from that luminary be as they are observed and experienced.

In short, for the production of these phenomena as they exist on the earth, if the sun be one million miles from the earth, its surface—hence, its diameter—must be very much less than that due to a diameter of 882,000 miles, for were it not, its light, heat and attraction would necessarily, be much greater that they are.

As appears by the formula called, "Arithmetical proportion," a proportion consists of four terms; the first and fourth terms are called the extremes, the second and third, the means.

The proportion sought, according to this formula, should be as follows :

As 93 millions of miles—being the estimated distance of the sun from the earth—is to 882,000 miles—the estimated diameter at that distance—so is 1 million miles—its real distance—to 9483⅞ miles, the assumed diameter of the sun at the latter distance.

Multiply the means, namely, 882,000, by 1,000,000. The product of the means is 882,000,000,000.

Divide the above product of the means by one of the extremes, namely, by 93 millions, thus:

93,000,000)882,000,000,000(9483⅛

Divide the product of the means by the *other* extreme, by 9483⅛, thus:

9483⅛)882,000,000,000(93,000,000

According to this formula the proportionate diameter of the sun, one million miles from the earth, is 9483⅛ miles.

On this basis, what should be the intensity of the light and heat of the sun as manifested on the earth ?

As a foundation for an estimate of these phenomena the following axioms are quoted, and relied on :

Axiom 1. "Light increases as the square of the distance of the illuminated from the opaque body." (The same law governs the increase of heat.)

Axiom 2. "To find the surface of a sphere we multiply the circumference by the diameter."

According to which axiom the surfaces of the larger and smaller spheres—representing suns—are found as follows :

Diameter of smaller sun.	Diameter of larger sun.
9483.8740	882.000
3	3
Circum. 28,431.6250	Circum. 26,460,000
9,483⅛	882,000

26,983,116,550,468) 23,337,720,000,000,000

Quotient, 8649

It will be found, according to these figures, that there are 8649 surfaces of the smaller sun contained on the surface of the larger sun.

Light increases as "the square of the distance."

The squares of the distances of the larger and smaller suns from the earth will shown the comparative intensity of their light and heat as effects of *distance*.

Distance of smaller sun from the earth 1,000,000 miles. Square of this distance, 1,000,000 times 1,000,000 equals 1,000,000,000,000.

Distance of larger sun from the earth, 93,000,000 miles. Square of this distance as follows : 93,000,000 times 93,-000,000 equals 8,649,000,000,000,000,000.

Square of smaller sun, 1,000,000,000. Square of lesser distance in favor of smaller sun, 8649. Excess of surface of larger sun over smaller sun, 8649.

So the account between the two suns, in loss and gain, —resulting from surface and distance—of light and heat, stands as follows :

Excess of surface—and consequently of light and heat —of a sun having a diameter of 882,000 miles over a sun having a diameter of 9483 miles, is equal to 8649.

Gain of a sun having a diameter of 9483 by lesser distance squared, namely, at the distance of 1 million miles from the earth, 8649.

Loss of a sun with a diameter of 882,000 miles, by reason of greater distance squared, namely, at 93 millions of miles from the earth, 8649.

Gain of such a sun at the latter distance, by reason of greater surface, 8649.

From this it will be observed that while the larger sun loses by greater distance from the earth, it gains over the smaller sun by larger surface. And while the smaller sun loses by lesser surface, it gains by reason of its shorter distance from the earth than that of the larger sun.

It will be seen by the estimate above made that this loss and gain of the two suns, in light and heat, are in the *same* proportion.

Hence, the conclusion is deduced that a sun distant one million miles from the earth, having a diameter of 9483 miles, will manifest the same intensity of light and heat, on the earth as would the sun were it 93 millions of miles from the earth, having a diameter of 882,000 miles.

In regard to the diameter of the sun, at the distance of. one million miles from the earth, it may be said :

If the sun distant 93 million miles from the earth, having a diameter 882,000 miles, be of such proportion as to surface and distance that the light and heat of that body manifested on the earth would result from such proportionate surface and distance, and this will not be denied by astronomers, then a sun one million miles from the earth, will have a diameter of 9483⅜ miles, for two reasons:

1. Because this is the only diameter at such distance which is proportional to the distance and diameter of the sun of astronomy. That it is proportional may be found by consulting the *first* figuring calculation made in this chapter.

2. It is the only diameter which will correspond with the light and heat of the astronomical sun. That they do correspond, and are proportional, may be observed by the figuring just made.

This will be at once apprehended ; if not, the proper calculations will prove this assertion, as to the correctness of the diameter, to be true.

Hence, the deduction is made that the real diameter of ·the sun at the distance of one million miles from the earth is 9483⅜ miles.

In regard to heat it may be said that while the sum total of it emitted from any heated body should be determined by the mass or capacity of such body, and not by the surface, yet that portion of such heat radiated from such mass in any *one* direction should be measured, so to speak, by the extent of that surface from which the heat were radiated in such direction and received by the body to which it was communicated.

For example, suppose a square bar of iron 12 inches long and 1 inch square were heated. While the sum total of this heat would, abstractly considered, depend on the mass of the heated bar, yet it would be radiated from 50 square inches, this being the number of square inches of the surface of the mass.

At an end of the bar, only $\frac{1}{50}$ part of the heat of the mass would be emitted, because such end would contain but the $\frac{1}{50}$ part of the whole surface, hence, were a body presented in close proximity to an end of such bar it would receive from it but the $\frac{1}{50}$ part of the heat radiated from the bar, the remainder, namely, the $\frac{49}{50}$ parts of such heat would pass in all other directions into space.

For analogous reasons that portion of the *surface* of the the sun which presents itself both immediately and mediately in the direction of the earth—and not the 'whole mass of the sun—would be the measure of the intensity of the heat received by the earth from it, the balance of such radiated heat passing into space, as in the case of the heated bar.

The contents or mass of any designated sphere is as the *cube* of its diameter, while the *surface* of such sphere is in much less proportion, namely, as the *square* of its diameter.

From these facts it might be thought that while the square of the diameter would properly measure the sun's light, the cube of the diameter should be the measure of its heat.

The foregoing illustration of the action of heat radiated from the iron bar is intended to show by analogy that the heat received by the earth from the sun would emanate from that part of the sun's surface presented to it, and that the amount of the heat so received would depend on the extent of the surface in such direction.

Hence that the *square*—and not the cube—of the diameter would measure the intensity of the sun's *heat*, as well as its light, as manifested on the earth.

ATTRACTION.

The relative attractions of spheres of different masses, of the same density, are as follows :

"Let us compare the attractions of two spheres of the "same material of which the diameter of the one is double "that of the other, the larger will have eight times the "bulk, and therefore eight times the mass of the smaller.

"But against this is the disadvantage that a particle on "its surface is twice as far from its centre as in the case of "the smaller sphere, which causes a diminution of one- "fourth. Consequently it will attract such a particle with "double the force that the smaller sphere will; that is, "the *attractions* are directly as the *diameters* of the spheres "if the densities are equal."—*Newcomb's Popular Astronomy.*

In order to find the attraction of the sun at one million miles from the earth, and having a diameter of 9483⅞ miles, in comparison with the attraction of a sun 93 millions of miles from the earth, and having a diameter of 882,000 miles, let it be supposed, for example, that this latter named sun be moved from this distance to that of one million miles from the earth.

In such position the comparative attractions of the larger and smaller suns—due to mass—according to the law last quoted, would be as their diameters.

At one million miles from the earth the relative attractions of each of these sun spheres for the earth—and reciprocally the earth's for each of them—would be also in proportion to such diameters, as such results would be in accordance with this law, as due to comparative mass, subject, however, to such change as might be required by the inverse square of the respective distances.

Calculation will show that the larger sun has contained in it 93 diameters of the smaller sun.

Then the smaller sun has but the $\frac{1}{93}$ part of the attractions of the larger, due to mass alone, which these diameters represent.

Again, suppose the larger sun removed back to its astronomical distance, namely, 93 millions of miles from the earth.

On computation of the attraction of the larger sun at this remote distance, what will be the comparative attractions of these two spheres for each other—and in relative proportion for the earth—as a result of difference of *distance* of such two spheres from the earth ?

This may be determined by Newton's law of attraction, which is as follows :

"Every particle of matter in the universe attracts every "other particle with a force directly as their masses, and "inversely as the square of the distance which separates "them."

This remote distance of the larger sun from the earth, in comparison with lesser distance of the smaller sun from the earth, is as 1 to 93.

The square of this greater distance of the larger sun from the earth is 93 times 93, equals 8649. The inverse square of this distance is 93, as 93 times 93 equals 8649.

According to this law of decrease of attraction, as a consequence of the increase of distance "which separates them," the larger sun, at 93 millions of miles from the earth, would have—as a result of such greater distance—but the $\frac{1}{93}$ part of the attraction for the smaller sun it had for such smaller sun, when the larger sun was one million miles from the earth.

But the smaller sun at one million miles from the earth would have, at such distance—as has been shown—the

5

$\frac{1}{93}$ part of the attraction that the larger sun had when it was at such near distance to the earth.

While, as has just been shown, the larger sun removed from one million miles from the earth, to 93 millions of miles from it, would there have but the $\frac{1}{93}$ part of the attraction it had when it was one million miles from the earth, having thus lost by distance in the remote position what it had formerly gained by mass when in the near position.

Then a sun at a distance of one million miles from the earth, and having a diameter of 9483⅓ miles, and a sun 93 millions of miles from the earth, and having a diameter of 882,000 miles, exert equal attractions on each other, and exert the same relative attractions on the earth, in proportion to the relative diameter of the latter body to each of the two suns, one of which is supposed to be 93 millions, and the other 1 million miles from it.

Hence the attraction of the sun exerted on the earth, at the distance of 1 million miles from it, and having a diameter of 9483⅓ miles, is equal to that of a sun 93 millions of miles from the earth, having a diameter of 882,000 miles.

Note.—It might be asked would the smaller and nearer sun have the *same* apparent size to the *eye* as the larger and more distant sun ? Light increases and decreases as the square of the distance. The sun proper has 8649 times the surface of the sun of the new theory. Were the larger sun moved from 1 million miles from the earth to 93 millions of miles from it, in such case the light of the sun would decrease as the square of the distance, namely, 93 times 93 equals 8649. Then the sun at the remote position would have but the $\frac{1}{8649}$ part of the light visible at 1 million miles, but the smaller sun has the $\frac{1}{8649}$ part of the surface of the larger, while the light of the larger sun moved to the remote position has diminished 8649 times what it had at the supposed near position. The remote and the near sun would then appear to the eye to be of the *same* size.

THE SOURCE OF THE SUN'S HEAT.

While the sun has a constant and undiminished supply of heat, the source of it is unknown.

There are many theories intended to solve this problem ;

One of these is that heat, lost to the sun by radiation, is replenished by matter in space falling to the sun, thus keeping up the solar fire.

Another is that this regular supply is continued by the recombination, or burning of compound gases at the surface of the sun.

Another theory maintains that the sun is a cold body, a magnet, and that magnetism is the agency by which heat is generated, and communicated to the earth.

None of these theories appear based on sufficient evidence, and none of them are generally accepted by scientists.

The contraction of the sun has been generally accepted as the most plausible source of supply of the sun's heat.

The theory of which is that as its heat is radiated into space the sun cools, and, as a consequence, contracts, thus generating heat to supply that lost to it by radiation.

This theory is founded on the conclusion that in the long ago the bulk of the sun was so vast as to have been commensurate with a very large portion of space, and that it is now—as it always had been—decreasing in bulk by its constant cooling, and consequent contraction.

The process supposed, is that the sun cools by throwing off its heat, in all directions, into space, and that, as a result of such cooling, contraction of its bulk takes place.

As a result of such contraction, *more* heat is generated than is lost to it by such cooling, and that these processes are continually going on, whereby a regular and undiminished supply of heat is furnished to the earth.

The philosophy of the theory is based on the fact, shown by experiment, that when a gas cools, it contracts and that "the heat generated by the contraction *exceeds* "that which it had to lose in order to produce the con- "traction."

It is also a fact that, "When the mass of gas is so far "contracted that it begins to solidify or liquify this action "ceases to hold, and further contraction is a *cooling* pro- "cess."

There is a weak point in this theory as to this contraction and cooling being the cause of the sun's heat, in this, to-wit :

It is apparently taken for granted that after each contraction of the sun it *remains* in the gaseous state, or, in other words, that the gas in contracting as it cools does not begin to become liquid or solid, and the fact that it does *not* is essential to the tenability of the theory, for according to the experiment, just quoted, did it "begin to solidify or liquify" there would be no heat produced in excess of that lost, in order to produce the contraction, hence the supply of heat to the earth would gradually diminish, for when the gas begins to liquify or solidify *this* is a *cooling* process.

We have nothing better than mere *conjecture* by which to determine whether the gas, so contracted, has begun to solidify or liquify or not, for it is one of the conclusions of astronomers that "We cannot yet say whether the sun has, or has not, begun to solidify or liquify in his interior."—*Newcomb's Popular Astronomy.*

The want of this essential knowledge renders this contraction theory untenable.

The theory evolved by the present writer is that the heat of the sun results from its motion—about the earth and betwixt the tropics—in passing through a resisting medium which occupies space, and that such heat is so generated in accordance with "the mechanical theory of heat."

That motion may be a cause of heat, was the opinion of the celebrated Locke He says :

"Heat is a very brisk agitation of the insensible parts "of the object, which produces in us that sensation, from "which we denominate the object *hot;* so that what in our "sensation is *heat* in the object is nothing but *motion.—* *Locke's Ele. Nat. Phil. C.* 11.

This view is confirmed by experiment, for we are told by Prof. Newcomb (Pop. Ast. p. 395,) that "It is now es- "tablished that heat is only a certain form of motion; that "hot air differs from cold air only in a more rapid vibra- "tion of its molecules, and that it communicates its heat "to other bodies simply by striking them with its mole- "cules and thus setting their molecules in vibration.

"Consequently, if a body moves rapidly through the "air, the impact of the air upon it ought to heat it, just as "warm air would, even though the air itself were cold."

In regard to the degree of heat produced by the rapid motion of air acting on a body at rest, Prof. Newcomb says :

"Let us apply this principle to the case of the meteor- "oids. The earth moves in its orbit at the rate of 98,000 "feet per second, and if it met a meteor at rest, our atmos- "phere would strike it with this velocity.

"By the rule we have given for the rise of temperature, "98,000 divided by 125 equals 784 times 784 equals "600,000 degrees nearly."

In speaking of the constant radiation of heat from the sun, he says:

"But it is now known that heat cannot be produced ex- "cept by the expenditure of force, actual or potential, in "some of its forms, and it is also known that the available "supply of force is necessarily limited. * * * * *

"Hence, this radiation cannot go on forever, unless the "force expended in producing the heat be returned to the "sun in some form."

If the distance and the mass of the sun be reduced, as before shown, and the argument in Chapter IV be sufficient to prove that the earth does not rotate, then the revolution of the sun about the earth is necessary to produce the phenomena of day and night.

As a consequence of such revolution, and having the velocity herein given it, the molecules of the sun would be put in agitation, or vibration, become hot, because the friction, caused by such rapid motion, between them and the ether "which occupies space," the sun striking them would be a cause of heat, in accordance with the mechanical theory of heat, just before quoted.

Were it conceded that such is the cause of the sun's heat, it would not follow that the supply of such *force* would be "necessarily limited," as stated in the quotation just made from Prof. Newcomb.

For in such case, what *force* can be either lost or diminished in the sense of exhausting the supply by the motion of the sun generating its heat by its striking the molecules of the ether, and what force is thus expended either "actual or potential ?"

The *force* generating the heat is the sun in its motion striking the ether, and it will be conceded that neither the sun, the ether nor any motion are diminished thereby.

As matter is indestructable, combustion could not diminish its mass ; as an effect of heat there would result changes of condition, form and consistency, but the quantity of matter would be the same.

Those portions of it on or near the surface of the sun, being thus nearest the source of its heat, would become dilated by it, their gravity would be thus diminished ; being furthest from the axis of the sun's motion, the centrif-

ugal force, as a consequence of this, would act more pow-
erfully on such matter than on any other on the sun.

As a result of the operation of these two causes, the
molecules of such matter would be thrown outward, and
by operation of a well known law of motion, they would
also move *onward* with a velocity in common with that of
the sun.

Being thus removed from the source of the most intense heat, they would there cool and condense, thus their
gravity would be increased. By the friction of the molecules of this projected matter with those of the ether in
space against which it would impinge, the centrifugal force
would be destroyed, while the gravity of this condensed
matter by reason of such condensation would be *then*, as a
force, in excess of the centrifugal force ; this force of increased gravitation would start such matter towards the
sun ; its momentum, according to the law governing falling bodies, and its *increased* gravity, caused by the condensation as just mentioned, would overcome the centrifugal
force, and such matter would fall to the sun to be dissolved
by the heat, and again thrown outward and onward as before, and thus the corona, as observed by aid of the telescope, might rationally be accounted for.

Another result of the constant throwing outward of this
incandescent gas or vapor by the centrifugal force, and its
return to the sun by increased gravity, as just surmised,
would possibly be a constant disturbance of the liquid
level of the surface of the sun, which is observed by the
depression of the photosphere shown through the telescope,
and such disturbance of level would, aided by the intense heat, rationally account of the currents, whirlpools
or spots, and the rush of matter so disturbed to and from
the sun's equator, poles and, in fact, in all directions, as is
revealed by the telescope.

Is there any law which could act *sui generis*, by which

the intensely heated condition of all matter observed at
the sun could be generated ?

Considering that friction acts on all matter in motion as
a constant force tending to stop it, and taking into view the
quieting agency of gravity, and its great power always acting in the direction of the sun's centre, and the constant
perturbation of all the matter observable at the sun, and
when it is further considered that the heat of the sun is always radiated and never diminished—like the contents of
Fortunatus' purse, always expended, yet always full—can
we conceive it possible that these observed facts and phenomena can forever continue on any other hypothesis than
that they result from the action of some force *external* to
the sun ?

The theory that the sun's heat is generated by its rapid
motion through the ether in space might rationally account for all these phenomena as effects of the mechanical
theory of heat, for on that principle—and with a less velocity than that of the sun, if it revolves—meteors are
ignited.

To the tenability of this theory, two facts are essential,
namely :

1. That a resisting medium occupies space.
2. That the sun moves rapidly through it.

As auxillary to the consideration of the first of these
two questions, we quote Sir John Herschel's estimate of
the height of the atmosphere. He says,

"Laying out of consideration all nice questions as to the
"probable existence of a definite limit to the atmosphere
"beyond which there is absolutely and rigorously no air,
"it is clear that for all practical purposes, we may speak
"of these regions which are more distant above the earth's
"surface than the one hundredth part of its diameter as
"*void* of air."

The one hundredth part of the diameter of the earth is

less than 80 miles, beyond this, for all practical purposes, is void of air.

As to the height which meteors have been observed above the earth, Prof. Newcomb says:

"The general result was that they were first seen at an "average height of 75 miles, and disappeared at the height "of 55 miles. There was no positive evidence that any "meteor commenced (that is to say, appeared ignited, for "otherwise they could not have been seen,) at a height "*much* greater than 100 miles."

Then the inference is that there was "positive evidence" that they appeared as *much* as 100 miles above the earth.

M. Biot, in his Elementary Treatise on Astronomy, in speaking of meteors, says:

"They are, in fact, ignited or inflamed bodies which "suddenly present themselves in the atmosphere, and "move with a velocity equal, in some instances to that of "the sun* in its orbit, or 20 miles in a second."

Without citing authorities or making quotations, it will be found that philosophers have come to the following conclusions, namely:

1. At 10 miles above the earth the atmosphere has lost ⅓ of its density.

2. At 45 miles it has become so attenuated as to be incapable of reflecting the rays of the sun.

3. At 80 miles above the earth space is void of an atmosphere for all practical purposes.

Then it must follow that at 100 miles above the earth it has no existence for any practical purposes, but at the distance of 100 miles above the earth meteors have been observed in an ignited state, and in that condition "enter the atmosphere."

It is the general conclusion of scientists that the ignition of meteors is an effect of the mechanical theory of heat.

---

* M. Biot, being a Copernican, means the earth's, not the sun's, orbit.

What resisting medium, then, does the moving meteor strike 100 miles above the earth, and beyond the *possible* limits of the atmosphere?

In confirmation of the almost necessary conclusion, that a resisting medium occupies space and that meteors in passing through which, with the velocity just quoted from Biot, would certainly be ignited, we quote Sir William Herschel in regard to the fact that the periods of comets are constantly diminishing.   He says:

"This is evidently the effect which would be produced "by a resistance experienced by the comet from a very rare "etherial medium pervading the regions in which it moves; "for such resistance, by diminishing its actual velocity, "would also diminish its centrifugal force, and thus give. "the sun more power over it to draw it nearer.   Accord- "ingly, no other mode of accounting for the phenomenon "in question appearing, this is the solution proposed by "Ericke, and generally received."—*Herschel's Astronomy, Title "Comets."*

Again, he says,

"A singular circumstance has been remarked respecting "the change of dimensions of the comet of Ericke in its "progress to and retreat from the sun, namely, that the "real diameter of the visible nebulosity undergoes a rapid "contraction as it approaches, and an equally rapid dila- "tion as it recedes from the sun.   M. Valz, who, among "others, had noticed this fact, has accounted for it by sup- "posing a real compression or condensation of volume, ow- "ing to the pressure of an ethereal medium growing more "dense in the sun's neighborhood."—*Idem.*

Did the foregoing, confirming the hypothesis that an ether occupies space, require authority for support, the names of Newton, Descartes, Kant and Euler, and their authority may be invoked, for each of these was of that opinion, as also is such opinion entertained by the more eminent modern scientists.

Prof. Tyndall virtually announces the general conclusion of modern science on this point, when he says :

"Luminous ether fills stellar space, it makes the uni-
"verse a whole, and renders possible the intercommunica-
"tion of light and energy between star and star. But the
"subtle substance penetrates further, it surrounds the very
"atoms of solid and liquid substances."

Again, the sun appears in an agitated and intensely hot state, and shows on a gigantic scale all the phenomena observed of a fire on a hearth.

If vibrations be necessary to the generation of heat, in fact, the cause of it, and there be no ether in space, vibrations—that is to say, heat—is not possible, for there would be nothing to be moved.

If vibrations—that is to say, heat—be transmitted from the sun to the earth and its atmosphere, then space cannot be a void, for vibrations mean, necessarily, an agitation of *matter.*

As to the existence of the other fact, before mentioned, namely the progressive motion of the sun in space, it may be said that while such motion is a *sine qua non* to the tenability of the theory, herein advocated, as to the source of the sun's heat, it must be admitted that there is no direct or indubitable evidence either of this motion of the sun or of its immobility.

But facts and phenomena do exist of such motion— and nothing contradictory of it— from which a *reasonable* conclusion may be deduced that the sun has a progressive motion about the earth and that it oscillates between the tropics.

For the present* the following may be cited,

1. The appearance of the motion of the sun is as plain-

---

* The strongest evidence of such motion will be considered in the next chapter.

ly observable as that of the moon and planets, and these bodies are known to have proper motions.

2. The earth does not appear to rotate, and the evidence relied on by astronomers as proving such motion may be reasonably doubted, as shown in Chapter IV.

3. All known phenomena—including the seasons—could exist, in accordance with natural law, did the sun so move ; this will not be denied, and this is much in support of such theory.

4. At the distance of the sun—found by the new measurement—such motion would not violate the laws of light, heat, or attraction as hereinbefore shown, nor would such motion be impossible by the action of centrifugal force, as will be hereinafter shown.

5. The appearance of ignition of the sun, would be such as it would have, did it move rapidly through a resisting medium. This may be rationally inferred from the fact that meteors are observed in an ignited state, moving through a resisting medium, above the possible limits of the atmosphere, and moving with a much less velocity than that herein assigned to the motion of the sun.

6. The velocity of 250,000 miles an hour—and this the sun must have according to the new measurement of its distance if it move about the earth in 24 hours— would be calculated to cause the vertical and cyclonic condition observable on the sun, and it would be difficult to conceive of anything else that would, and if this be true it *per se* amounts to some evidence that the sun does so move.

7. The contraction theory of the cause of the sun's heat may be regarded as untenable for the reason that it will not account for the phenomena of vast masses of liquid fire, for the immense volume of heated gases and vapors revealed to observation by the telescope, showing such to be the normal condition of the sun. By that theory contraction of the entire body of the sun takes place as a

consequence of its cooling. Such cooling results from
the radiation of its heat into space; this cools the
sun; this of course negatives the idea that the sun
retains the heat so generated, for did it do so, cooling would not take place, and this is as necessary to
the contraction as cooling is to the theory, while as evidence that cooling does not take place the telescope shows
there storms of fire and dashing billows of flame always
observed, which shows such to be its normal condition,
which negatives the theory that the sun is ever cool.

8. The observed fact that the sun cuts the meredians
and the parallels *obliquely*, which while it would do this
on the Copernican theory of the earth's progressive motion,
it would also do so on the theory herein advocated.

Because did the sun revolve about the earth and also oscillate between the tropics as it is observed to do, the direc
tion of such motions would be as though around and coincident with the direction of the threads of a spiral or screw,
hence it will be perceived—did the body on which the
threads of the screw were formed represent the axis of the
earth—that motion in the direction of the threads could
not be coincident in direction with that of the parallels,
as these are at right angles to the direction of the earth's
axis, hence consistent with such motions of the sun it
should cut the parallels and meridians obliquely to them,
and to the earth's axis as it is observed to do.

To the theory of the source of the sun's heat herein
advocated the following objections—as also many others
—may be made.

1. It might be objected that the velocity of the sun
namely, 250,000 miles an hour,* moving through a resist-

---

* Newton calculated the velocity of the comet of 1680 to be 880,000 miles an
"hour. * * * * * * * * Brydone calculated that the velocity of a
"comet which he observed at Palermo in 1770 was at the rate of two millions
"and a half of miles in an hour." Parker's Philosophy, page 374.

ing medium, would result in dissipating the mass of the sun into vapor.

To expect this it is necessary to assume that a mass of matter so immense as that constituting the sun could be heated to the intensity required for such result, which would imply that before such intense heat was generated, radiation of heat from it would *not* take place.

While on the contrary such radiation should be expected before such intense heat were generated from such a large body as the sun, as it would heat more slowly than a small body, owing to its great bulk, and would radiate heat constantly and gradually before the heat became intense and thus vaporization of the whole mass would be prevented.

For example as the sun became heated by its motion through the ether, dilitation of the ether immediately surrounding it would result, as an effect of such heating, and such dilitation would diminish the frictional capacity which had, supposably, existed between the molecules of the ether, and those of the moving sun before such dilitation had taken place.

The diminution of the heat as an effect of radiation would be immediately followed by the cooling, hence the contraction of the ether, and thereby it would become more dense—and thus of greater frictional capacity—than immediately before the heat had been so reduced.

This greater density would result in increasing the heat as an effect of the increased frictional capacity of the more dense ether.

This increased heat would again expand the ether, and so on, when a law of proportion as to its expansion and contraction would bring about uniformity in the vibrations of the ether, hence in the quantity of the heat thus generated.

By virtue of this necessary process, the accumulation of

sufficient heat in the body of the sun, required to vaporize it, would be impossible.

---

2. Another objection to the theory of the sun's progressive motion may be that a very great and a very destructive centrifugal force would be generated by its velocity in revolving about the earth, namely that of 250,000 miles an hour.

From the following argument it does not appear probable that this velocity would generate a centrifugal force sufficient to throw the surface matter of the sun beyond the range of its attraction, and on this hypothesis only, could such objection have weight. however great such force might be.

Were this force great enough to do this, the planet Mercury, being the largest body nearest the sun, would receive such matter thus thrown off as an effect of its attraction.

And as Mercury is yet the smallest of the primary planets, hence its yet small volume furnishes no evidence of any increase of bulk from this or from any other cause.

The reason for the opinion that the centrifugal force would not—in the case supposed—throw matter from the sun beyond the range of its attraction is partly founded on the correspondence between the centrifugal force which would be generated by the rotation of the earth, in case it did rotate, and that of the sun did it revolve about the earth in the orbit, and with the velocity assigned it.

The energy of the centrifugal force in any case depends upon the velocity of the body in motion, and upon the extent of the deviation of such motion from a straight line, and the measure of such force is proportional to such velocity and such deviation.

The distance of the sun from the earth, by the new measurement, is one million miles. The circumference of its orbit, about the earth, is then 6,000,000 of miles, approximately, while that of the earth is about 24,000 miles.

So the orbit in which the sun would revolve about the earth is 250 times the circumference of the *orbit* of the earth—so to speak— in its rotation,

The velocity of the earth on its axis is about 1,000 miles an hour, while the velocity of the sun, in the case supposed, is 250,000 miles an hour.

Thus while the velocity of the sun, in revolving in its orbit about the earth would be 250 times that of the earth's, did it rotate on its axis, yet the angle of the tangent to the *curve*—so to speak—of the sun's orbit is 250 *less* than that of the earth's, that is to say, the curve of the sun's orbit is but the $\frac{1}{250}$ of the curve of the earth's surface.

So the centrifugal force due to the sun's velocity, in excess of that of the earth's, would be correspondingly diminished as a consequence of the *curve* of the sun's orbit being *exactly* as much *less* than that of the curve of the earth's surface as the velocity of its revolution in the orbit is in excess of that of the earth's rotating on its axis.

Thus the nearer approximation to a straight line in the path of the sun preventing that centrifugal force due to its velocity, while the greater curve, and corresponding tangent, of the orbit in which the earth would rotate would be calculated to generate a greater centrifugal force than that of the sun had the earth the velocity of the sun, which velocity the earth not having this is compensated by its greater curve, while correspondingly the lesser curve of the sun's orbit is compensated by its greater velocity, resulting in the centrifugal force in each being the same.

The centrifugal force of the earth's rotation has been calculated, and found to be the $\frac{1}{289}$ part of terrestrial gravity.

The conclusion from the foregoing estimate is that the earth could rotate without its surface matter being thrown, by the centrifugal force generated thereby, beyond its attraction, as such force would be the $\frac{1}{289}$ part of terrestrial gravity.

Hence that the same result would take place with the light, heated vapors and gases which observation of the sun shows as forming the corona, resulting, perhaps, at least partly, as an effect of the centrifugal force, generated by the velocity, and the curve of the orbit, of the sun.

---

3. Another objection, to the theory of the sun's revolution about the earth, may be made, which will be considered in a subsequent part of the present section.

As preliminary, the following is quoted from M. Biot, he says :

"These phenomena (namely, the phases of the planet "Venus) represent Venus as a kind of moon revolving "around the sun, and illuminated by its light. All obser- "vations confirm this fact."

As the planet Mercury revolves about the sun and has phases, the foregoing quoted conclusion that Venus is a satellite of the sun applies as well to Mercury.

As satellites of the sun that primary would carry the orbits of these planets with it about the earth did it so revolve, as all planets having satellites, namely Mars, Jupiter, Saturn, Uranus, and Neptune carry the orbits of their respective satellites with them, in the revolutions of these planets about the sun and earth.

Were the distances of Venus and Mercury reduced in

6

proportion to the reduced distance of the sun by the new measurement, namely, by using 93 as a divisor of the astronomical distance, such distances would not preclude the possibility of the sun so carrying the orbits of these two planets.

On such supposition, a little reflection will show that all the phenomena observed in connection with these planets would be the same ;

And the more readily will this be perceived when it is considered that the earth is not included within the orbits of either Mercury or Venus.

Did the sun carry the orbits of these planets about the earth, they would move with the velocity of the sun.

In such case, an objection might be made that their heat should be that of the sun, generated as that of the sun, according to the mechanical theory of heat, and that observation does not show the phenomenon of heat as observed on the sun.

In the direction of an answer to this supposed objection it may be said :

Were two globes, one of phosphorus, and the other of asbestos, for example, revolved rapidly in the air, the velocity which would ignite the phosphorus would scarcely heat the asbestos globe at all.

And of what matter these planets are constituted, hence their combustibility, or absence of this quality, is unknown, and for this reason the absence of evidence of the same intensity of heat as that observed at the sun is not decisive that the orbits of these planets are not carried by the sun in its revolution about the earth.

Again, while the matter constituting these planets has not the agitated appearance that the matter of the sun is observed to have, yet these planets appear as they might were they incandescent, Venus being the brightest object in the heavens, except the sun, and Mercury not much less so.

And the fact that they have phases—from which it is inferred that they appear by reflected light—would not disprove the assertion that they may be incandescent, as an effect of their motion through a resisting medium as a result of the mechanical theory of heat.

For while the shadows rested on them—which is the cause of their phases—these portions of their surfaces would be darker than those where the shadows were not.

This would be the result of the contrast between a greater and a lesser intensity of light whether these planets were normally brightly illuminated or not.

That such might be the case may be inferred from the fact that while the planets and stars are bright bodies and some of them very bright, yet they are made, by the more intensely bright light of the sun so dark as to be totally invisible in the day, with the exception of Venus, which is so intensely bright that when the air is very clear it may, sometimes, be seen in the day time by the naked eye, and under the same conditions, it will cast a shadow at night, thus showing a greater brightness than any other bodies in the heavens except the sun and moon.

Did Venus receive its intense light from the sun—and it is the general opinion that it does so—it could only do so in violation of a cardinal law governing the intensity of light radiated or reflected on a dark body, which law is, that light increases or decreases, as the square of the distance separating the luminous from the opaque body.

According to astronomical estimate, Venus is thirty-one millions of miles further from the sun than is Murcury.

As results of this law and this fact, did Venus receive its light from the sun, it should manifest to us less light than does Mercury, in the proportion that the square of the distance of Venus from the sun is greater than that of Mercury.

But the fact is that Venus, at a greater distance from

the sun—namely thirty one millions of miles—than that of Mercury, is the brighter.

Then as its more luminious appearance is contradictory of this law, it must be self luminous, which it would be did the sun carry its orbit about the earth.

As an answer to this argument it may be said that the square of Venus' greater distance from the *sun* than that of Mercury is also the square of its lesser distance from the *earth*, and that it is just as much nearer the earth as it is further from the sun, and that it is this lesser distance from the earth that gives it a brighter appearance than Mercury.

This conclusion, however, must assume that Venus has the *same* intensity of light thrown upon it from the sun as Mercury has.

But this is taking for granted that which the squares of the distances between Mercury and Venus and the sun prohibits, namely, that the intensity of the light of the sun in both these planets were the *same.*

Were this the case, *then* the somewhat brighter appearance of Venus than Mercury might result from the earth being nearer Venus than it is to Mercury, but not otherwise.

As to the heat of Mercury, it is the conclusion of astronomers that "The heat of this planet is so great that "water cannot exist there except in a state of vapor, and "metals would be melted."—*Parker's Phil.*

These facts are attributed to the heat of the sun ; were this so, its light should be proportionately intense, yet its light is less than that of Venus, which is much further from the sun than is Mercury.

Whereas, were it regarded as an immense meteor moving with its orbit and revolving with the sun about the earth, its heat might also thus be accounted for.

On the transit of Venus and Mercury they have the appearance of dark bodies, but this may result from the *more* intense light and heat of the sun giving them this appearance and does not prove that they are *per se* cold and dark bodies.

For a like reason, we cannot see the planets and stars in daylight, namely, because of the *more* intense light of the sun.

In regard to Venus, we are told by astronomers that "The heat and light at Venus are nearly double what they "are at the earth."   And again, "Spots are also sometimes "seen on its surface like those of the sun."—*Parker's Phil.*

As reasonably explanatory of these phenomena, we may say that M. Faye, of Paris, and Father Secchi, of Rome, made careful observation of the phenomena manifested at the sun, and speak at great length and minuteness of the flames, the incandescence, the vertical motions, the dashing billows of inflamed matter, the whirlpools in the currents of liquid fire there observed on the entire apparent surface of the sun.

Father Secchi says, "These whirlpools *constitute* the "spots."

M. Faye says, "The *whirlpools* of the sun are like those "of the earth of all dimensions from the scarcely visible "pores to the *emormous spots* which we see from time to "time."

The spots, then, on the sun are whirlpools of liquid fire.

When it it considered that Venus has the appearance it would have were it incandescent ; that by estimation its heat is double that at the earth ; that it is brighter than Mercury, which is so much nearer the sun than Venus ; that Venus has "spots like those of the sun"; that the spots of the sun are generated by the most intense heat ; taking all these facts into consideration, would they not amount to some evidence that these phenomena may result from the mechanical theory of heat, caused by the motion of Venus with the sun in revolving about the earth ?

While it has been shown in this chapter that the heat of the sun, with its reduced distance and bulk, would be the same as before such reductions, it may yet be objected that the measure of such heat is not that due if generated by the mechanical theory.

According to this theory, a body moving through the air 125 feet in a second generates one degree of heat, as before stated.

The sun, according to the theory of its motion, would have a velocity—in round numbers—of 69 miles in a second, which is equivalent to 364,320 feet in a second. This divided by 125 is equal to 2914, the square of which is 8,491,396, which are the degrees of heat at the sun. that is, if the ether has the density of the air.

What portion of this heat would come to the earth cannot be known, for it is conceded by astronomers that a very large, but not accurately known, portion of the sun's heat is absorbed by its atmosphere.

La Place estimates this to be as much as eleven-twelfths of the whole.

What portion of it ceases to exist in space by the friction—which, while it causes heat, eventually stops motion—of the molecules of the ether caused by their rubbing against each other, thus stopping their vibrations—which vibration is heat—cannot be known.

Again, there is no means of knowing that the ether, through which the sun is herein supposed to move, is as dense as our atmosphere or not.

Should it not be, the ratio of the rise of temperature of bodies passing through the air, namely, one degree per second for every 125 feet of distance, would be no guide or criterion by which the intensity of the heat of the sun received at the earth could be estimated.

Father Secchi, of Rome,—who is regarded as an authority in philosophical questions—estimated the heat at the sun to be several millions of degrees, which approximately confirms the estimate here made, according to the mechanical theory of heat, were the ether of the same density as the air.

In concluding the consideration of the source of the sun's heat, it may be said that neither the contraction and cooling theory hereinbefore discussed, nor any other as to the source of the sun's heat—regarded by scientists as philosophic—will rationally account for the fiery phenemena observed on the sun.

2. The ignition of meteors above the possible limits of the atmosphere; the diminution of the velocities of comets as they proceed in the direction of the sun, and their dilation as they recede from such direction ; the compression and dilation—as they thus proceed and recede —of their neuclei, prove two facts, namely : First, That a resisting medium occupies space, and Second, That a body rapidly moving in such medium is thereby ignited.

3. That the theory of the motion of the sun in this medium with the velocity herein assigned to that luminary, will sufficiently account for its appearance of ignition, and rationally explain the phenomena observed.

4. As no other theory but the sun's motion in such resisting medium will explain the phenomena observed, and the theory herein maintained will, this amounts to evidence *per se* that it does so move.

CHAPTER IV.

DOES THE EARTH ROTATE?

SECTION I.

M. Biot was one of the most able, as also one of the most distinguished, of modern astronomers.

His work entitled, "Elements of Astronomy," continues to be a text-book in colleges.

In that work he presents the evidence and the argument on which astronomers rely in proof of the Copernican system of astronomy.

As a necessary part of his system, Copernicus maintains that the earth rotates.

The following constitutes the evidence of such rotation, namely:

1. The velocity of the beats of the pendulum is diminished at the equator.

2. The earth bulges, or is protuberant, in the region of the equator, and flattened at the poles.

3. The fact that gravity augments, on proceeding from the equator, in the direction of the poles.

4. The fact that bodies, from great heights, fall *east* of a vertical line.

5. The spheroidal form of the planets Jupiter, Mars and Saturn, so similar to that of the earth, constitutes evidence, from analogy, of the rotation of the earth.

6. Considering the distance and magnitude of the sun, the argumentative deduction is made that it is "more simple to suppose" that the earth rotates, than that the sun revolves about it.

Astronomers maintain that the four enumerated facts

first mentioned are caused by centrifugal force, and thence deduce the conclusion that the rotation of the earth generates such force.

On the hypothesis that the centrifugal force diminishes gravity, the existence of these facts amount to *prima facie* evidence of such rotation, but they do not constitute conclusive evidence of it, for the reason that the mere existence of these facts do not necessarily preclude the possibility of their being the result of causes independent of such rotation.

That such causes exist, will be attempted to be shown in the subsequent sections of the present chapter.

The estimated distance, and magnitude of the sun—in comparison with the magnitude of the earth—is regarded by astronomers as an argument on behalf of the theory of the rotation of the earth, even of greater force than the foregoing enumerated facts.

The force of this argument may be correctly estimated by the reader after perusal of the second chapter, wherein the method of the measurement of the distance of the sun is explained, and whereby it's true—and diminished—distance has been found.

Section II.—The Cause of a Slower Beat of the Pendulum at the Equator, Considered.

Is it philosophical to feign a hypothesis, namely, the centrifugal force, to account for a slower beat of the pendulum at the equator, than elsewhere, when there exist two known causes for this, both of which act on the pendulum in diminution of gravity ?

Indeed, these causes necessarily act on every particle of matter there, and with greater force in diminution of gravity than at any other part of the earth's surface.

For example, the earth is a spheroid.   At the equator its radius is about 13¾ miles longer than at the poles.

Gravity diminishes or increases as the ratio of the distance.

As the distance from the equator to the earth's centre is the greatest, according to this law all bodies must have less gravity at the equator than at the poles, and as a consequence the pendulum at the equator should beat slower, as its gravity is less than in the direction of the poles.

Again, a cause for a slower beat of a pendulum at the equator may be deduced from the fact that as the equator is 13¾ miles further from the centre of the earth than are the po'es, it would follow as a result of that fact, that the equator is 13¾ miles nearer to such celestial bodies as are constantly over the equator, as also such as never diverge at any great angle from the vertical line over it.

The sun is never at a greater angle to such line than 23½°, nor the moon at a greater angle than 29½°.

The greater number of "the high host of stars" are also within these ranges.

The attractions of the sun and moon on the earth are greatest within the equatorial region.

This is proved by the fact that the tides are highest there.   Then these attractions must act in diminution of of gravity there to a greater degree than elsewhere on the earth.

In order to illustrate the law governing the attraction of the sun—which law is common to the attractions of all bodies—at the equator and elsewhere on the earth, and the effect of this law, we quote from a standard authority on Physics, as follows, illustrating the tides:

"The point of the earth's surface which is nearest the "sun will gravitate towards it more, and the remoter parts "less than the centre inversely as the square of their re-"spective distances.   The point A (vertically under the "sun at the equator), at the earth's surface, therefore tends

"away from the centre, and in such case the fluid surface
"must rise.   This effect must be diminished in proportion
"to the distance from the point A in any direction, and at
"the North and South poles of the earth 90° distant it
"ceases."

This quotation establishes two facts, namely :

First.—That at the point of the earth's surface nearest
the sun the attraction of that body on matter on the earth's
surface is greatest.

Second.—That such attraction diminishes in proportion
to the distance from such point.

Two causes for a slower beat of the pendulum at the
equator have been shown in this section, namely :

1. Gravity would be diminished there, because the distance to the earth's centre is greatest there.

2. The sun's attraction is greater there in diminution
of gravity than elsewhere on the earth.

Did not the causes mentioned, namely, 1 and 2, diminish the gravity of the pendulum at the equator and thus
cause it to beat slower there, as it is observed to do, it
would be an inexplicable anomally.

If, on the other hand, these causes are sufficient to account for the slower beat of the pendulum there, such
slower beat is no evidence of the rotation of the earth, as
it is said to be.

SECTION III.—THE CAUSE OF THE PROTUBERANCE OF THE EARTH AT THE EQUATOR, CONSIDERED.

The protuberance or bulge of the earth at the equator
is, by astronomers, attributed to the centrifugal force generated by the rotation of the earth.

Because, by reason of the greater distance of the equator
from the axis of the earth, this force would be greater
there than elsewhere on the earth.

It is a fact, philosophically determined by geologists and conceded by scientists generally, that the earth was, at one period in the inconceivably distant past, in a hot, fluid condition.

M. Biot, in Sec. 114 of his work, says, "It appears from "a great number of facts in natural history that this state "has really existed, and that the earth was once a *fluid.*"

And that as it cooled, slowly, in the long lapse of ages a solid crust—now constituting the surface of the earth—was thereby formed.

With the exception of this crust, evidence exists justifying the opinion—generally entertained by scientists—that that the earth remains in this hot condition.

Did the earth rotate, and were the matter constituting it once a fluid, or even in a plastic state, the bulge of the earth around the equator might possibly be the result of the centrifugal force generated by such rotation, because centrifugal force changes the direction of gravity, as experiment proves.

The fact that this bulge exists at the equator does not *per se* prove it to be caused by the centrifugal force ; for the reason that another force has always acted strongly on the earth, at the equator, in restraint of gravity. Such force acts there with greater energy than elsewhere at the earth's surface, just as the centrifugal force would.

Is it equal there to that centrifugal force, to the action of which such bulge is attributed by astronomers ?

Before attempting to ascertain the extent of this power, counteracting gravity, it may be pertinent to remark, *en passant*, that the rotation of the earth, even in the estimation of astronomers, is but a *theory*, hence it belongs, rather, to the domain of rationalistic logic, so to say, than to that higher and indubitable logic called demonstration.

Hence, it should not be forgotten by the reader that to doubt the tenability of this theory—this speculation—does not involve the absurdity of questioning the truth of a mathematical demonstration.

That the attractions of the celestial bodies were acting forces at the equator and were such as would be calculated to swell the earthy matter there, when in a fluid state, to the form of a protuberance, may be rationally inferred from the following facts, and deductions from them, namely :

1. The celestial attractions do counteract gravity, notably the attractions of the sun and moon. This fact is proved by the existence of the tides, which, it is universally admitted, are caused by the attraction of the sun and moon, whereby the gravity of the waters of the oceans are partly overcome.

2. That the celestial attraction is greater at the *equator*, is shown by the quotation from "Natural Philosophy," made in the last preceding section.

It also appears, in that quotation, that the attraction of the sun diminishes in proportion to our recession from the equator, in any direction.

These instances are sufficient to prove :

1. That the attractions of the celestial bodies do counteract the force of gravity.

2. That such attraction, and consequent force counteracting gravity, is greater at the equator than elsewhere on the earth.

3. That such attraction diminishes in proportion to the distance we recede from the equator.

What then is the amount of the celestial attraction—the measure of its energy—so far as such is exerted on the *bulge* of the earth at the equator in opposition to gravity ?

In order to make an approximate estimate of the extent of this, so acting, suppose there are two spheres, *each* having a diameter of 7912 miles, such being the mean diameter of the earth at 45°; this latitude being about midway betwixt the equator and the poles of the earth.

Suppose a solid ring of matter 13¾* miles in thickness to be slipped over the surface of one of these spheres, co-incident with the line of its circumference. Suppose this ring to slope from the circumference line, gradually North and South, as the bulge of the earth slopes.

This latter body would then fairly represent the sphe-roid of the earth, and its diameter at the equator.

The celestial attractions—principally of the sun and moon—exerted on this ring, or protuberance, may be, at least approximately, estimated as follows :

The radius of the earth at the equator is 3963 miles, and its diameter there is, in course, 7926 miles. If the length of this radial line, namely 3963 be divided by the thickness of the bulge at the equator, namely by 13¾ it will be found that the latter number is contained in the former 288¼ times, say, in round numbers 288 times.

And in course, conversely, if 288 be multiplied by 13¾ the radius of the body having this ring or protuberance will be that of the radius of the earth at the equator, namely 3963 miles, if the omitted fractions be added.

As the bulge increases the radius at the equator 13¾ miles, it would increase the diameter twice this number, namely, 27½ miles.

If then the equatorial diameter of the earth, namely, 7926 be divided by 27½ the quotient will be found to be 288, and a fraction.

As "the attractions of spheres, of the same density, are as their diameters," and as the *equatorial* diameter of the sphere—changed to a spheroid by the bulge—is the $\frac{1}{288}$ part greater than that of the other sphere, whose diameter is supposed to be taken at the 45th parallel, this *excess* of

---

* This estimate of the thickness of the bulge may not be exact, but is at least a close approximation to it, as it has been estimated by some as being 13 miles thick, by others, 13½ miles. Measurements of the radii of the earth have been made at different latitudes, and the *exact* measurements of these have been found very difficult, and these different measurements, made by astronomers, are found to differ from each other.

equatorial diameter, namely, that of the bulge, must have $\frac{1}{288}$ in excess of attraction, at the equator, over that exerted at the equator of the sphere having the diameter of 7912, and having *no* bulge.

The bulge or protuberance of the earth at the equator then exerts the $\frac{1}{288}$ part *more* of the attraction of gravitation than would be exerted at the equator of the earth did it not have any bulge.

When it is considered that attraction between bodies are equal and reciprocal *pro tanto* it will follow that the celestial attractions exerted on this bulge will be the $\frac{1}{288}$ part of the attraction they would exert on the whole earth did it not have such bulge.

By the same process of mathematical deduction as that which finds that the attraction of spheres are as their diameters, the *radii* of such different diameters will also show the comparative gravity, at the *equator*, of each of such two spheres.

Hence, as the radius of the earth at the *equator* is the $\frac{1}{288}$ part greater than that of the sphere whose diameter is supposed to be taken at 45°, it would seem to follow that the bulge of the earth at the equator should approximately represent the $\frac{1}{288}$ part of terrestrial gravity.

Let us now enquire how this celestial attraction exerted at the equator compares with the centrifugal force supposed to be exerted there.

On the assumption of the truth of the hypothesis that the earth rotates, and that the centrifugal force diminishes terrestrial gravity, astronomers have mathematically determined the ratio of the increase of this force proportionate to any given increase in the velocity of the earth's rotation, as also the effects of such increased velocity on the matter of the earth.

In accordance with such calculation they find, and assert, that did the earth rotate 17 times faster than its present asserted velocity, the centrifugal force generated thereby would be that of 17x17 equals 289.

They thence deduce the conclusion that centrifugal force would "equal terrestrial gravity, and as a consequence, bodies at the equator would weigh *nothing*.

From this calculation it follows that the present velocity, given the earth, must generate a centrifugal force at the equator equal to the $\frac{1}{289}$ part of terrestrial gravity.

But—as before shown—the celestial attraction exerted on the equator is *also* equal to about the $\frac{1}{289}$ part of terrestrial gravity. It is then *equal* to the centrifugal force.

When it is considered that there is no mathematical demonstration of the earth's rotation ; that there is no direct, or positive proof of it ; that there is no *sensible* evidence of it, for as M. Biot observes in speaking of the hypotheses of the diurnal motion of the celestial bodies or that of the earth, that "appearances them-"selves present absolutely nothing by which it can be de-"cided which of the two hypotheses is true."

And when it is further considered that the chief evidence of such rotation is that furnished by the centrifugal force and by analogy, from which a deductive inference is drawn that the earth rotates, is it philosophical to account for the bulge of the earth as the effect of the centrifugal force, when it may be rationally considered as the result of the celestial attractions acting on matter when in a fluid or plastic state and most strongly at the equator ?

For if the deduction before made as to the energy or amount of this attraction be not technically mathematical, the result found by the calculation is approximately correct.

Besides, even were it not so calculated, the sufficiency of this attraction to raise the plastic matter of the earth at the equator above the level due to gravity, may be *rationally* conceded when it is considered that Newton proved the precession of the equinoxes to be the attractions of the sun and moon on the bulge of the earth, whereby the earth is given a slow motion of revolution, its pole moving around the pole of the heavens.

This attraction causes both the motion and the variances and differing velocities of such motion. For it is found by observation and calculation that when the sun and moon are, at the same time, on opposite sides of the equator, and on opposite sides of the axis of the earth, variances in the uniformity of the precession of the equinoxes are observed.

It is also found that the precession is greater when the sun is nearest the earth.

This change in precession must result from the increased attraction of the sun on the bulge of the earth, in consequence of the sun's greater proximity to it.

This motion of the earth, so caused, gives us such conceptions of the enormous power exerted by the attractions of the sun and moon on the bulge of the earth, as to rationally induce the conclusion that it were sufficient to create the bulge, at that period in the earth's history, when the matter constituting it was in a fluid or else plastic state, even were it impossible by any calculation to learn the *exact* amount of this attracting energy, adverse to gravity, exerted by it on the bulge of the earth.

An additional argument, as to the sufficiency of this power to form the bulge, in the long ago, may be deduced from the fact that the waters of the ocean, seas, &c. cover about two-thirds of the surface of the earth.

The attractions of the sun and moon at the bulge are sufficient to raise these waters, in the form of tides, and thus *pro tanto* overcome their gravity.

Hence the bulge of the earth at the equator—as also a slower beat of the pendulum there—independent of the previous calculation, may rationally be inferred as the ef-

7

fects of that celestial attraction which causes the precession of the equinoxes, and which counteracts gravity as
observed in connection with the tides.*

The *modus operandi* by which the celestial attraction
formed the bulge cannot be known, but in theory it may
be conceived of; for example, the density of matter would
be lessened and its gravity be diminished by the expan-
sive agency of heat, at the period when the matter of the
earth was in a hot fluid or plastic state.

The cooling of such matter would first take place where
it was furthest from the centre of the heated mass, namely,
at the surface of the earth.

Such result would be in accordance with natural law,
as manifested by observed phenomena.

That portion of the heated mass constituting the bulge
is nearest the earth's surface. The matter forming the
bulge is nearest the attractions of the sun and moon. In
consequence of such greater proximity, the attractions of these bodies on it would be greater, as it is observed that the motion of precession of the equinoxes is increased when the sun is nearest the earth.

The attractions of the sun and moon—greatest at the
equator—would raise this fluid, or plastic, mass above the
normal level due to its gravity. The tides show such
result.

The constant attraction of these bodies, on possibly a
viscid mass, in a direction opposite to that of gravity,
would prevent the ebb of the mass so raised.

This outer portion of the heated mass, even were it
melted to a fluid, on continuous cooling during the lapse
of ages, would become a crust immovable, and solid, thus
forming and constituting the bulge or protuberance belted
around the earth at the equator.

---

* This argument assumes that the sun and moon existed either antecedent to,
or else were contemporaneous with the first existence of the earth. These implied assumptions may be regarded as one of the defects of the argument.   -

Section IV.—The Cause of the Augmentation of Gravity in the Direction of the Poles, Considered.

In the last two preceding sections the attempt is made to show, argumentatively, that the celestial attractions will rationally account for a slower beat of the pendulum at the equator, and, indeed, for the observed diminution of the gravity of all bodies there, and will also account for the bulge of the earth.

In the present section it will be shown—by aid of a mathematical formula—that while the centrifugal force will change the *direction* of gravity, it will not diminish the *quantum* of the gravitating force.

If this be proved, great force will be added to the conclusiveness of the hypothesis that the phenomena just mentioned are effects of the celestial attractions; and, *pro tanto*, will negative the assumption that they are caused by the centrifugal force, and will, to that extent, weaken the theory of the earth's rotation, which these phenomena are supposed to support.

In this connection, the first question is, what is the direction of the centrifugal force?

The following facts, which have the authority of axioms, are an answer to the question:

"1· The direction of the centrifugal force, is the direc-"tion of a line, tangent to the curve, which a body de-"scribes in its motion.

"2· The tangent of a circle is a line which touches the "circumference without cutting it when lengthened at "either end."

That the centrifugal force will not diminish gravity is shown in connection with Fig. 6, together with Fig. 8, shown in Section 5.

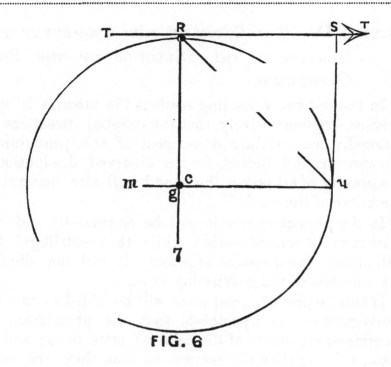

FIG. 6

This circle will be supposed to represent the earth.   C its centre.   R g its radius from R.   This radius will also be the direction of gravity from R.   At R the line T T is tangent to this circle, as it *cuts* the circumference nowhere. This line, then, shows the *direction* of this force at R— according to axiom 1—as due to the earth's rotation. The quantity of this is represented by the length of the line R S. *

In order to show that the centrifugal force will not diminish gravity, let it be supposed that this force were sufficient to carry the body R from R to S, along the tangent line T T, were there no other force acting upon it.

* As a deduction from astronomical calculation, this centrifugal force is only the 1–289th part of terrestrial gravity.   Then the line R S is very much too long to correctly represent its *quantity* in comparison with the length of the line of gravity R g. yet that the centrifugal force will not diminish gravity, may be as correctly shown—and much more plainly—by the line R S, as if it were drawn proportionately, will not be doubted,

Again, suppose the gravitating force would carry the body R to g, along the line R g, were there no other than this gravitating force acting upon it.

But again, suppose *both* these forces to meet at **R** and to act on the body **R**. What will then be the direction and the quantity of these two forces?

In order to determine this, from R draw R S, thus representing the quantity and direction of the centrifugal force R S. From R draw R g, representing in quantity and direction the gravitating force R g. From g draw the line g n, parallel to R S, and from n draw the line n S, parallel to R g. These lines will thus form a parallelogram. Then draw the line R n, and this will constitute the diagonal of this parallelogram, and the direction and quality of these two forces—being their resultant—will be measured by the diagonal R n.

This conclusion is a deduction from the following mathematically demonstrated proposition, namely :

"If two forces be measured in direction and quantity by "the sides of a parallelogram, a single force measured in "direction and quantity, by the diagonal of such parallel-"ogram, will be the resultant and mechanical equivalent "of such two forces."

According to this proposition, the direction and quantity of the gravitating and centrifugal forces, R g, R S, will be measured by the length of the diagonal R n.

What proportion in quantity, does the resultant R n bear to the normally gravitating force R g?

Where a force acts in a certain direction, say in the direction of the resultant R n, it is sometimes required to know what portion of such force acts at some angle—say, for example, that of R g—not coincident in direction with that of the force itself.

There is a formula laid down in mathematical works to ascertain the quantity of this angular force.

This formula is applicable here, for the reason that while the force of R n does not move on a line coincident with that of gravity namely R g, yet it exerts a force in the direction of gravity—namely that of the line R g—for if it did not, the body R could never reach the earth, vertically, as it does.

This method and its results may be shown in connection with Fig, 8, in Section 5. The quantity and direction of the force are shown by the diagonal line a 4, in that figure.

It is required to know what portion of this force acts coincident in direction with the line a c 5. To ascertain this, from 4 draw the line 4 c, perpendicular to the line a c 5, and the line a c will represent the portion of the force a 4, which will act coincident with the line a c 5.

While it is true that the force a 4 may be resolved into the two forces, namely, a c, a D, yet it is equally true—according to the formula as explained in mathematical works—that the portion of the force a 4, which acts in the direction a 5, is measured by the line a c.

Examine Fig. 6 in connection with Fig. 8, and draw n c m perpendicular to R c 7—Fig. 6—and that portion of the resultant force R n, which acts coincident with the line R c 7, will be the line R c.

This result is in exact accordance with the formula explained in connection with Fig. 8. Then the line R c represents that part of the force R n which acts on the line of gravity.

But the line R c represents the original gravity of the body R, *before* its force combined with the centrifugal force ; then the centrifugal force does not cause *any* diminution of the force of gravity.

In confirmation of the correctness of this conclusion, reference is here made to Fig. 18, in Section 5.

The cannon ball is fired in a horizontal direction, which is also the direction the ball R moves from R on the tan-

gent line T T ; yet the projectile force of the cannon ball
—analogous to the centrifugal force at R, Fig. 6—does not
diminish nor restrain gravity, for the reason that the cannon ball reaches the earth's surface at the same time that
another ball does, which is dropped from the same height
as the other, and at the same time.   This is the result of
experiment.

By operation of both the formula before mentioned and
of the experiment just referred to, two facts are established, and one as certainly as the other, namely :

1. That the centrifugal force will not diminish gravity.

2. That the combination of the centrifugal with gravity
will change the direction of the latter.

Then, as the centrifugal force does not diminish gravity, for that reason, it cannot be true—as is asserted by
astronomers—that the centrifugal force diminishes the
gravity of the pendulum at the equator.

Neither can it be true that gravity is augmented in the
direction of the poles, *because* the centrifugal force would
be diminished in the same direction.   For while it is true
that the centrifugal force would be diminished in that direction, yet, as it does not diminish gravity, (did it exist,)
gravity cannot be augmented by the diminution of the centrifugal force.   Hence, it cannot be true, as asserted by M.
Biot in Sec. 117 of his work, that :

"The increase of gravity from the equator to the poles
"is, *therefore*, a new proof of the rotary motion of the
"earth."

This quotation embodies the opinion of all astronomers.
As the diminution of the centrifugal force in the direction
of the poles is discharged of the responsibility of augmen-
ting gravity in the same direction, as shown by the argument in connection with Fig. 6, what is the cause of such
augmentation ?

In Section 3 it is shown that did the centrifugal force exist, the attractions of the celestial bodies would be equal to it at the *equator.*

M. Biot, in speaking of the centrifugal force and its diminution, says :

"It is also easy to prove by mechanical principles that "this diminution is proportional to the square of the sine "of the latitude like that of attraction."

Will the equatorial celestial attractions, diminished by greater distance, in the direction of the poles, philosophically account for the observed augmentation of gravity in the latter direction ?

That such would be the result, may be inferred from the last quotation from M. Biot.

In order to show the extent of the diminution of the celestial attraction—as represented by the sun—in the direction of the poles, draw a circle—or a semicircle—say, for example, having a diameter of six inches. The three-inch radius of it will represent the centrifugal force at the equator.

As shown in Section 3, the celestial attractions are *equal* at the equator to the centrifugal force.

Then, if this radial line be continued from the equator vertically an *equal* distance, namely, three inches, it will correctly measure the extent of the celestial attractions at the equator in comparison with the centrifugal force there, each being represented by the figure 3.

Imagine the sun to be at the extremity of this line, furthest from the equator. Draw a straight line from such extremity, namely, from the sun, to 45°.

Attraction is as the square of the distance ; then the attraction at the equator is 9, as this is the square of the distance, 3.

As the celestial attraction and the centrifugal force at the equator have been shown—see Sec. 3—to be equal, that force is also 9 at the equator.

The attractions of bodies decrease, inversely as the square of the distance increases.

. The line from the equator to the sun is 3 inches ; from the sun to the 45th parallel it is 4⅜ inches. The celestial attractions at the equator are the *square* of this, 3, namely, 9. Apply the above law, as to the inverse ratio, to ascertain the sun's attraction at 45°, and it will be found that the greater distance has reduced this attraction of 9 at the equator to $4\frac{37}{100}$ at 45°.

By reason of the sun's decreased attraction at 45° it has lost 4⅜ of the attraction of 9, it has at the equator.* Calculation will show this.

At 45°—on the circle mentioned—draw the sine of the latitude and it will be found to be two inches, and $\frac{1}{12}$ inch, and a small fraction.

The centrifugal force diminishes as the square of the sine of the latitude. The square of $2\frac{1}{12}$ is equal to 4⅜. Then the centrifugal force has lost 4⅜ of the 9 it had at the equator.

Then the diminution of the celestial attraction in the direction of the poles, and the diminution of the centrifugal force in the same direction are equal to each other.

Then the augmentation of gravity in the direction of the poles may be attributed to the diminution of celestial attraction in the same direction. And as this force equals the centrifugal, the observed phenomena of the augmentation of gravity in the direction of the poles, would still be observed, even though the centrifugal force had *no* existence.

M. Biot, in Sec. 116 of his work, says, in speaking of the diminution of the centrifugal force and the acceleration of the pendulum in the direction of the poles :

. "Now by carrying the same pendulum to different parts "of the earth, it has been found to move faster as we de-

---

* When the sun is vertical over the equator, the reduced height of the tide at 45°, shows an equal diminution.

"part from the equator; and this acceleration, the laws
"of which have been determined with great exactness, is
"in fact proportional to the square of the sine of the lat-
"itude as *required* by the rotation of the globe."

That is to say, the pendulum has just such acceleration
as it would have did the earth rotate, and the exact acceleration—no more and no less—is due to the centrifugal
force generated by such rotation.

Did this centrifugal force exist, and did it diminish
gravity, the following questions would require answer,
namely:

If this augmentation of gravity is the effect of the diminished centrifugal force *alone*—and it is, according to
the last quotation from M. Biot—what becomes of the
celestial attractions, as an equal agency in the production
of this result?

If the celestial attractions—equal to the centrifugal
force—produce this result, what becomes of the centrifugal force?

If both these forces, by their joint action, augment
gravity to the extent observed, how can it be correctly
attributed to the centrifugal force *alone*, as M. Biot does
in the last quotation, for this force is that *exactly sic* due
to the earth's rotation?

Did both of these *equal* forces contribute to the result,
their *double* force of retardation should augment gravity
but one-half that observed, which shows that but *one* of
these forces is causing the phenomenon.

Which, of these two forces, is the cause of the phenomenon?

The attraction of the sun and moon at the equator, and
the diminution of this attraction as we recede from the
equator, are *facts*; the precession of the equinoxes, and the
tides, and their action, prove them to be facts.

The rotation of the earth—and its progeny, the centrifugal force—is a hypothesis.

In order to show that such hypothesis is not sustained, the three following conclusions are reached, namely :

1. The attractions of the sun and moon are *equal* at the equator, to the centrifugal force, on the admission that such force exists. This is shown in Section 3.

2. In connection with Fig. 6, it appears that the centrifugal force does not diminish gravity. This is also argumentatively shown in connection with Fig. 18.

3. In the present section it appears that the celestial attractions diminish in the direction of the poles, in the *same* ratio as would the centrifugal force, did it exist as an effect of the rotation of the earth.

In view of these facts is the following conclusion, reached by M. Biot, logical, namely :

"The increase of gravity from the equator to the pole is "*therefore* a new proof of the rotary motion of *the earth*,"

We can conceive of a centrifugal—or some other—force, say, for example, 289 times that due to the rotation of the earth, which would throw a body beyond the attraction of the earth. In such case, gravity would be absorbed, or destroyed by the centrifugal force, and in such case the two forces, did they exist, would be coincident in direction.

For this reason it would be impossible to construct a parallelogram representing the *two* forces, hence no diagonal representing a resultant of such two forces would be possible.

With the centrifugal force, represented as being the $\frac{1}{289}$ part of terrestrial gravity,—as in this case—different laws act on the gravitating body than would were gravity to act *coincident* in direction with the centrifugal force.

For example, (see Fig. 6,) were a body in motion to impinge on a body at rest, or not moving in the direction of the attacking body, or moving slower in the same direction, the attacking party loses a portion of its force in overcoming the inertia of the body impinged upon. According to this law, what the former loses the latter gains. Then the centrifugal force R S,—Fig. 6,—expends a portion of its energy in deflecting the body R in the direction R n, and instead of the centrifugal force acting in diminution of gravity, it *adds* a portion of its force to the gravity of R, in the direction R n.

But this addition of force is not seen in the result, as R reaches the earth in the *same time* it would did it fall on the line R g, and thus operated by gravity alone. This is shown to be the fact by Fig. 18.

The cause of this identity in result may be explained as follows :

Measurement will show that the line R n is longer than R g, hence the ball, in falling to the earth on the line R n, traverses a *greater* space in the *same* time than it would did it fall along the line R g, by gravity alone. To fall through this greater space in the same time as the gravitating ball, requires power, it addition to the gravitating power. According to a mechanical law, what is gained in time or *space* is lost in power. The gain in space here is the difference in the lengths R g and R n ; this greater space on the line R n is gained by the ball on R n, as it falls to the earth in the *same* time as the merely gravitating ball R, in falling on the line R g. It requires a power in addition to the gravitating power to do this. This additional power is furnished by the centrifugal force, in accordance with the laws governing motion and force, as aforesaid.

This added power is exactly competent to bring about the result just mentioned. For were there more power received from the centrifugal force than necessary to bring the ball, moving on the line R n, to the earth at the same time as the ball falling by gravity alone, on the line R g, it would reach the earth sooner than the ball falling along R g. Were there less power it would not—

as it does—reach the earth at the *same* time as the gravitating ball.

From this analysis it will be perceived that instead of the centrifugal force diminishing gravity—as is asserted by some to be the fact—it, on the contrary, augments it, but, paradoxical as it appears, this augmentation is not apparent, for the reason that it is expended on the longer line R n.

By operation of the laws herein mentioned, in all cases where there is a *combination* of the gravitating and the centrifugal forces, and which produces a diagonal force, as a resultant of the actions of the two forces, it is easy to understand *why* the centrifugal force will not restrain or diminish gravity.

SECTION V.—WHY DO BODIES FALL—FROM GREAT HEIGHTS—EAST OF THE VERTICAL?

M. Biot, in Section 118 of his work says ;

"The rotation of the earth becomes further evident from "another remarkable phenomenon, namely the deviation "of bodies which fall from a great height.

"To understand this, let us imagine a heavy body "placed at a great distance from the earth's surface, at the "top of a high tower for example.

"If the earth is at rest the body will fall at the foot of "the tower in the direction of the vertical.

"But if the earth turns on itself, bodies which partake "of this motion will have, at the commencement of their "fall a velocity of rotation exceeding that of the base of "the tower on account of its greater distance from the axis "of motion.

"Thus when a body falls by the combined effect of this "horizontal motion and that of gravity it ought to meet "the earth a little in advance of the vertical in the direc-

"tion of the earth's motion, and consequently after its fall
"it will be a little distance *east* of the tower ; this is con-
"firmed by experiment."

Does this phenomenon prove the rotation of the earth ?
Two causes are assigned for this phenomenon, namely :

1. The longer axis of motion at the top, than at the
base of the tower.

2. The projectile, horizontal motion—of the earth's rotation—combining with the gravity of the descending
body. This conclusion, as to the cause of the phenomenon, is not logic in this, to wit :

The horizontal motion mentioned is that of the earth
on its axis.

This motion is inferred from the fact that the body—
say, for example, a heavy ball—falls from the top of a
tower *east* of it. This inferred or assumed horizontal motion is combined with the gravity of the descending ball.

From the combination of this *ideal* horizontal with the
actual gravitating force, the conclusion is deducted that
the fall of the ball *east* of the tower is *caused* by the combination, together with the greater velocity of the ball at the
top of the tower than at its base, by reason of its greater
distance from the axis of the earth. This conclusion is
defective for two reasons.

1. It virtually ignores the possibility of there being any
other cause for the phenomenon, than that inferred.

2. The rotation of the earth, which is the proposition
to be proved, is assumed in order to account for the fall of
the ball *east* of the tower, which position of the ball is relied on in proof of such assumption, which is not logical,
unless the ball *east* of the tower is the position it would
be found were such assumption a fact.

But aside from this criticism let us first enquire whether
this combination of gravity with a projectile force would
add any force of projection to the latter, before considering what effect a longer distance from the axis of motion

would have on the direction of the ball when falling from the top of a tower.

The effect of the combination just mentioned may be shown with Fig. 8, on the margin.

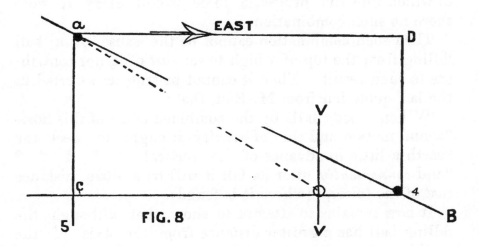

FIG. 8

"Here the force of projection would carry a ball on a "straight line—say for example eastward—from a to d, "while gravity would bring it to c.

"If the action of the air on the ball, retarding it, be "omitted, and these two forces alone prevailed, the ball "would move along a B, and would be carried to 4."

Measurement of the lines a d and c 4 will show them to be of the same lengths, also that the opposite lines are parallel. The ball is moved eastward from a to 4 by the combination of a c the gravitating force, with a d the horizontal projectile force.

These forces act on the ball falling from the top of the tower, if the earth rotates, and such *combination* contributes in projecting the ball *east* of the tower, according to M. Biot, as appears by the last quotation from him.

It will be observed that the ball at 4 is no further *east* than it would be were it driven eastward from a to d by the projectile force *alone,* seeing that d is the same distance *east* as is 4.

Then the combination of gravity with a horizontal projectile force will not carry a ball beyond the vertical line, to which line the projectile force would carry it were there no such combination.

Then such combination cannot be the cause of a ball falling from the top of a high tower *east* of it, nor contribute to such result. Then it cannot be true, as asserted in the last quotation from M. Biot, that :

"When a body falls by the combined effect of this hori-"zontal motion and that of gravity it ought to meet the "earth a little in advance of the vertical * * * * * "and *consequently* after its fall it will be a little distance *east* of the tower," (where it is found).

It now remains to attempt to show that although the falling ball has a greater distance from the axis of the earth's motion at the top of the tower than has the base of the tower, that such fact will not cause the ball to fall east of the tower.

1. It will be shown that when a projectile force is combined with gravity, the latter force is neither increased nor diminished by the combination.

2. It will be attempted to be shown that as the ball and tower have a velocity in common, the ball in falling from the top of the tower would have no longer axis of motion than that of the towor, and that the ball would not fall east of the tower because of its longer distance from the axis of motion.

The following is copied from a standard work on "Natural philosophy."

"In Fig. 18 is shown a cannon which is loaded with "ball and placed at the top of a tower at such a height as "to require just 3 seconds for another ball to descend perpendicularly.

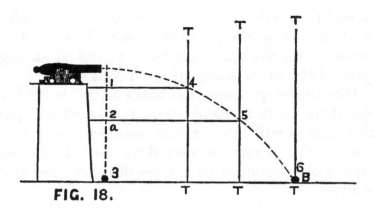

FIG. 18.

"Now suppose the cannon to be fired in a horizontal di-
"rection, and at the same instant the other ball to be
"dropped toward the ground.

"They will both reach the horizontal line 3 B at the
"same instant.

"In this figure the dotted line C a represents the verti-
"cal line of the falling ball. The dotted line C B is the
"curvilinear path of the projected ball.

"During the first second of time the ball reaches 1, the
next second, 2, and at the end of the third second it
"reaches the ground (3, B). Meanwhile that projected
"from the cannon reaches 4 while the other ball reaches 1,
"4, 1 being parallel to the horizon, showing that the two
"balls fall downward with the same velocity. During the
"next second the ball from the cannon reaches 5 while
"the other falls to 2, both having equal descent.

"During the third second the two balls will reach the
"ground on the horizontal line 3 B at the same time."

This *proves* that the force of gravity is neither increased
nor diminished by the force of projection.

In connection with Fig. 8 it has been proved that gravity does not add any force to that of projection. Then,
as gravity is thus out of the question, if any reliance is
placed on the fall of the ball *east* of the tower as evidence

8

of the earth's rotation it must be because the *greater* velocity of the ball at the *top* of the tower than at its base throws, or flings the ball from the top, and at an angle to that face of the tower along which it falls.

M. Biot in the quotation last made from his book, says:

"But if the earth turns on itself, bodies which partake "of this motion will have, at the *commencement* of their "fall, a velocity of rotation exceeding that of the base of "the tower on account of its greater distance from the axis "of motion."

Two or more bodies may have a motion in common having the *same* velocity in the direction of such motion, for example a boat and its passenger, a horse and its rider, &c.

According to the foregoing quotation the falling ball has "at the *commencement* of its fall a velocity of rotation "exceeding that of the base of the tower."

This, then, must be the point where the greatest projectile force acts on it. This then must be the point where it leaves the face of the tower and forms an angle to it ; but "bodies that partake of this motion," namely, the earth's rotation, "have a velocity of rotation" at the top of the tower "exceeding that of the base of the tower."

But the top of the *tower* is one of the "bodies which partake of this motion," and has at the "commencement" of the fall of the ball, a motion in common with it, the tower having at the commencement of the fall of the ball, the *same* length or distance from the axis of the earth's motion as the ball, must have the *same* velocity as the ball. In such case how can the ball get off the face of the tower, coincident with the face of which it is supposed to be at the instant it commenced its fall ?

The passenger having a motion in common with the boat could not get ahead of it, as an effect of such motion. Nor the rider ahead of his moving horse. To do this another and independent force would be necessary.

Equally so, and by operation of the same laws, neither could the ball at the "commencement" of its fall get beyond the face of the tower at the time it commenced to fall. Then no angle could be formed by the falling ball with the tower ; then it could not fall east of it and that it does so is *no* evidence of the earth's rotation.

To illustrate the action of the top of the tower and the ball at the instant it commenced falling, the two moving with a common velocity, suppose the cannon —Fig 18—were projected by some extraneous force, with the same velocity, in the same direction and at the same instant the ball were fired.

The ball and cannon would reach the earth at the same instant—if the resistance of the air be excluded—and the ball would not be projected from the cannon, because the ball and cannon would have *a velocity in common.* This result will be admitted. And for this reason the ball could not get in advance of the cannon, nor the falling ball in advance of the top of the tower. At the top of the tower, at the instant the ball commences to fall, the ball and the top of the tower have a velocity in common. How then can the ball get in advance of the moving tower? Whence comes the force that would project it "a little distance *east* of the tower ?" Certainly not from the earth's rotation.

The action of a pendulum may serve to show that the causes relied on by astronomers for the ball falling from a high tower east of it, will not produce this effect, for example .

Did a pendulum oscillate synchronously at the base of of a mountain, it will do so at its top, for everywhere the pendulum moves as though acted on by gravity alone.

Suppose such pendulum to oscillate East and West at the top of Pike's peak, or the Peak of Tenerife, for example. At the top of such peak the axis of the earth would be 3 or 4 miles further than to its base.

Gravity combined with the projectile force of the earth's rotation acting from the extremity of a longer axis of motion are said to cause the ball to fall east of the tower. As the pendulum oscillates east and west at the top of the peak it is acted on (1) by gravity, (2) by the projectile force due to the earth's rotation, (3) by a vastly longer axis of motion than that said to give velocity to the ball falling from the tower. Thus all the elements and conditions are present in the one case as in the other. Yet the action of the pendulum furnishes no evidence that it is acted on by the same forces as those said to move the ball from the line of the tower, for the reason that it beats synchonously at the top of the peak, as at its base. Yet did the earth's *rotation* move the ball *east* of the tower, the same force would move the pendulum ball east with greater velocity than it could move west, hence it could not beat sychronously.

Why is this the fact ? because if the earth does not rotate this phenomenon is that which will take place as an effect of terrestrial gravity.

Did the earth rotate, the pendulum itself would have a motion—namely, that of the earth's rotation—in common with its oscillating ball. As a consequence of this motion in common, even if the earth rotates it will act as observed, namely, as though acted upon by gravity alone.

A motion *in common*—resulting from the *same* forces as those acting on the pendulum ball at the top of the mountain—acts on ball and tower.

The fact that the ball is found east of the tower is evidence—by analogy to the pendulum experiment—that some cause *other* than rotation projects the falling ball east of the tower, for were it the effect of the earth's rotation, the pendulum ball placed at the top of the peak would have a longer distance from the axis of motion than would the ball at the top of the tower, and from the reason given for the fall of the ball east of the tower should be effected in a corresponding manner.

As an illustration of the effect produced where two or more bodies have a motion in common, suppose that a stationary railway train has a cannon fired on each end of the train simultaneously, each ball striking a target at the opposite end of the train at the same time.

Suppose again that the train be started, and to have *any* given velocity. Let the cannons be fired as before, and the respective balls will strike their respective targets at the same time, as before. The reason of this is, that while the ball fired from the rear has two motions, namely, that of the train combined with that caused by the explosive, yet the target on the front of the train is moving away from this fired ball with the same velocity as that given such ball by the motion of the train. Hence this ball has but the velocity—practically—given it by the explosive. The other ball moving from the front to the rear of the train has to contend against the velocity given it by the explosive, by the train moving in a direction opposite that of the ball, which *pro tanto* diminishes the velocity of the fired ball. But to compensate this loss the rear end of the train is moving towards this ball with the velocity of the train, which is exactly that which the ball has lost by the motion of the train opposite in direction to that given by the explosive; so the ball strikes the target on the rear end of the train impelled by the explosive alone, at the same time as did the other ball strike its target. This shows that two or more bodies having a motion in common act *towards each other* as though they had *no* motion. Will not this *law* apply equally to the action of the ball and tower?

Figs. 8 and 18 shows that the combination of a projectile force with gravity produces no other effect than changing the direction of these forces.

Suppose the curvellinear dotted line Fig. 18 to represent the curve of the earth's rotation.

Did the tower move to 4, by the earth's rotation, represented by the line "T T," the ball in virtue of the same projectile force, equal to that of the tower, would then, as before, be coincident with the line of the tower. But this force is combined with gravity, and while the ball proceeds to "T T" it falls to 4, the same laws acting upon it as those which brought the cannon ball to 4 in Fig. 18,

By operation of the same laws when the tower had moved in space, say for example, to the lines T 5—T 6, the ball would be on the face of the tower at 5, then at 6, and would fall to the earth on the line of the tower at b, and could not be, by operation of the laws astronomers rely on, "a little distance to the east of it."

If astronomers should maintain that the position of the ball where found east of the tower is that due to the causes which they say place it where found, and that its position is not aided by the action of any other cause, a plausible argument may serve to show that the position of the ball *east* of the tower is evidence that the earth does *not* rotate, instead of being—as it is regarded—the strongest evidence that it does.

It is a fact so well known that the friction and inertia of the air retards the velocity of bodies moving in it, that the many familiar examples of it may be omitted.

As the ball—Fig. 18—falls from the top of the tower to the earth in 3 seconds, according to the law governing falling bodies the tower will be 144 feet high.

Did the earth rotate, it would move 4392 feet in three seconds, estimating its radius at 4000 miles. Considering this short distance, the curve of the path of the earth would vary from a straight line but a few inches; hence, for illustration, this path in the short space of 4392 feet may be regarded as a straight or horizontal line.

Did the earth rotate, the ball in falling from the top of

a tower would be impelled by gravity and the projectile force resulting from the earth's rotation. Considering the experiment —Fig. 18—this will be conceded.

Did the earth rotate, the direction of the ball falling from the top of a tower would be that of an oblique line, that is to say, the ball would move neither coincident with the horizontal direction of the earth's motion, nor in the normal direction of gravity.

That the ball, in such case, would fall obliquely as just stated, may be admitted, when the explanations of Figs. 5, 8 and 18 are considered.

In such case the ball in falling would move coincident in direction with a hypothenuse line,—somewhat *curved* so to speak—of a triangle, of which the height of the tower would be the perpendicular, and the base that of the horizontal line.

This hypothenuse line would be, of course, longer than either of the two other lines. However, in any case, the length, as also the angle, of this oblique line would depend on the height of the tower, and the length of any given base line.

The atmosphere would move horizontal, coincident in direction with that of the earth and tower, and with an equal velocity, consequently the velocity of the tower would not be retarded, by the friction or inertia of the air.

But such would not be the case with a falling ball, because it would not move in the *same* direction as the atmosphere, as this together with the tower would move horizontally, while the falling ball would move oblique to such direction.

For this reason both the friction and inertia of the atmosphere, which the ball would meet on the oblique path of its motion, would be overcome by it.

The ball would necessarily expend a portion of its force to do this, and this as a consequence would result in a retardation of the velocity of the moving ball.

That such would be the case, may, perhaps, be shown by comparing the motion of the atmosphere and tower, coincident with the direction of the earth's motion, to a boat and the current by which it moves, having thus a motion in common, in direction of the course of a river, say for example.

In such case, as the boat floats along with the current, its motion is unimpeded by friction of the water on it, and it has no inertia of the water to overcome, because the boat and the water on which it floats have the same velocity, namely, a velocity in common.

But if the boatman turns his prow oblique to the direction of the current, and rows for the shore, in addition to the abstract force expended in reaching the shore, he must expend such force as may be necessary to overcome the friction and inertia of the water which resists his progress in the direction he is then going, as the water does not move in the oblique direction in which the boat is *then* moving.

The oblique direction of the ball in falling from the top of the tower—did the earth rotate—not being in the direction of the atmospheric current, is analogous to the boat illustration, and the falling ball would—as the boatman in the case of the boat—have to expend a portion of its force to overcome the friction and inertia of the atmospheric current, which does not move in the same direction as that of the falling ball.

In such case the ball would not move *so far* in the direction of its motion as it would have, had its motion not have been so resisted.

Calculation will show that while the tower, in common with the earth, will move 4392 feet on a nearly horizontal line in 3 seconds, a ball falling from the top of a tower 144 feet high, would in the same time reach the earth moving on an oblique line 5124 feet long.

The extent of the resistance of the air to the motion of this ball would be proportionate to the *difference* between the distance the ball falls along the oblique line mentioned, namely, 5124 feet, and the distance the earth and tower would move horizontally in the same time as the falling ball, namely 4392 feet in 3 seconds.

This difference is 732 feet, which is more than 5 times the height of the tower.

Let us now enquire as to the extent of the resistance of the air in the motion of bodies. The author, from whose standard work on "Natural Philosophy," Fig. 8 was copied, says :

"The force of projection would carry a ball from A to D —"see Fig. 8—while gravity would bring it to c.

"If these two forces alone prevailed the ball would pro-"ceed on the line a B to 4.

"But as the resistance of the air operates in direct op-"position to the force of projection, instead of reaching the "ground at 4 the ball will fall somewhere about the line V V."*

As the air would not be in motion in the oblique direc- tion the ball would move from the top of a tower carried along horizontally by a rotating earth, the ball moving on such oblique line would meet the resistance of the air along a line 732 feet long.

What this resistance may amount to may be estimated —in retarding the velocity of the falling ball—by the ex- tent of such retardation shown on Fig. 8.

A tower 144 feet high would be but the $\frac{1}{146.666}$ part of the radius of the earth.

Calculation on this basis will show that the velocity of of the top of the tower in excess of that of its base would be so *very* insignificant in comparison with the loss of ve-

---

* "In a note the author says, "It is calculated that the resistance of the air "to a cannon ball of two pounds weight, with the velocity of 2000 feet in a sec- "ond, is more than equivalent to 60 times the weight of the ball."

locity caused by the retardation of the air—which might be as much as one-sixth of the projectile force, as shown by Fig. 8—on a line 732 feet long, that it should bring the ball *west* of the tower.

Hence, the observed fall of a ball "a little distance *east* of the tower" does not prove the rotation of the earth as asserted by astronomers. * (See note).

M. Biot, in Sec. 498 of his work, says, speaking of the fall of bodies *east* of the tower.

"If we suppose the earth immovable these phenomena "admit of no explanation."

Yet, if the foregoing analysis of the laws he invokes, shows that these phenomena will not result from them the fact that they exist does not prove the rotation of the earth, even though these phenomena be inexplicable.

The present writer will now try his " 'prentice hand" in framing a theory which to him looks plausible although, possibly, it may not account for the phenomena just mentioned.

The capital law governing falling bodies is that they fall on vertical lines. In addition to this it appears that gravity exerts its force to bring falling bodies on such lines. This appears from the following quotation from a standard work on "Natural philosophy," namely:

"The force of gravity and the resistance of the air cause "projectiles to form a curve both in their ascent and de-

---

* The moon is always falling towards the earth, but gravity combined with the projectile force acting on the moon causes it to revolve about the earth. Seeing that it revolves about the earth its orbit would be, so to say, the circumference of a wheel, of which the earth's centre would be the nave. The moon would be most distant from the axis of motion. The ball is said to fall *east* of the tower, as an effect of the causes which act in connection with the moon's motion. But a motion of the moon thus caused is not observed, as astronomers assert, of the moon's motions, that "it is the theory of attraction which has furnished us with the knowledge and *measure* of these phenomena." Then the moon has *no* easterly motion, resulting from the mechanical principle which is said to give the ball a motion *east* of the tower, from whose top it falls.

This fact amounts to an argument adverse to the hypothesis that the ball falling *east* of the tower, is *caused* by the rotation of the earth, for were this true, the moon—also a falling ball—should be *east* of that position in space due to attraction alone, and it is *not* so found.

"scent; and in descending, their motion is gradually "changed from an oblique towards a perpendicular direc-"tion."

A few examples may serve to show the further action of this law. Suppose a ball rolling down an incline plane leaves it, midway the plane. At that instant, the future direction of the ball is governed by its momentum and its gravity combined.

Momentum alone, would move it along the plane. By gravity alone it would fall to the earth on a vertical line, from the point it leaves the plane. But after leaving the plane it is still acted upon by both of these forces. Acted on by its momentum, the ball cannot fall along a verticle line, and its gravity being unsupported it cannot move along the plane.

As a consequence of the action of these two forces it falls to the earth on a line oblique to the direction of the plane, and to that of gravity. Yet this ob'ique line will be more nearly coincident with a vertical line than is the inclination of the plane.

This example of the action of the law last quoted, shows that momentum has forced the ball *across* a vertical line along which it would have fallen when it left the plane, had gravity alone prevailed.

Another example will also show that momentum will cause a falling body to *cross* its normal vertical line. Suppose a heavy body to be suspended. If it be drawn aside from the vertical along which it hangs, and then left free to act, it will fall by gravity to such vertical, and impelled by the force of momentum thus acquired it will *cross* its vertical line and continue to cross and recross it, until friction and the resistance of the air stops its motion, when it will be found quiescent on the vertical, whence it was started.

These familiar examples prove that while a vertical is the normal line of falling bodies, as the gravitating force when unimpeded keeps them on such line, and draws them—so to speak—on such line when off it, as shown by the law last quoted, at the same time, when this force is overcome by momentum, it may, and does, under certain circumstances, force such body across such line, for example, in the instances just cited.

The application of such facts as those, together with some laws found in connection with falling bodies, form the foundation for the theory, as to the *cause* of the ball falling *east* of the tower, under the circumstances mentioned by M. Biot, in a previous quotation from him.

Some of the laws governing falling bodies are as follows :

1. "No two vertical lines can be parallel."

2. "No two bodies from different points can approach "the centre of a sphere in a parallel direction."

3. "The direction which a falling body approaches the "surface of the earth is called a vertical line."

We suppose the sides of the tower to be *parallel,* as buildings are generally constructed on such lines. In such case the faces, or sides, of the tower cannot be vertical. Because, according to the law just quoted, "No two vertical lines can be parallel" and, conversely, no two parallel lines can be vertical ; but—say, for example,—the eastern and western faces of the tower are parallel, then, for this reason they are not vertical.

Indeed, it is said that the suspended cords of a small weighing scale are not strictly parallel, but, owing to the shortness of these cords, their divergence from such parallelism is so slight that it is not perceptible.

As the faces of the tower are not vertical because they are parallel, the line of the tower would not be the direction a ball would *normally* fall from its top, because, according to law 3—just quoted—all bodies fall to the earth's surface on vertical lines.

In the instances cited, we find the ball on leaving the inclined plane, falls nearer a vertical line than it was when moving along the plane, and as near the line of the vertical as the momentum it has acquired will permit it.

The suspended ball moves towards its proper vertical, crosses and recrosses it by its gravity and momentum, and finally comes at rest coincident with it.

Why should a ball falling from the top of a high tower, and not on a vertical line—as it is not, as has just been shown—form an exception to the action of the laws quoted in this connection, and to the examples illustrating their action ?

Should not these laws act on it, and the result of their action be analogous to that which resulted in the cases cited ?

Should not the ball, in falling from the top of the tower, and not on any vertical line, *cross* its proper vertical, impelled by its gravity and momentum, as did the balls in the examples given, impelled by the same forces ?

If the ball, in falling from the top of a high tower, did not form an exception to the laws governing *all other* falling bodies, hence, act by some unknown force independent of such laws, it will approach its proper vertical at an *acute angle* to the line of the tower—because the vertical, and the line of the tower would vary very slightly from parallelism—being acted on by the laws governing the direction of the motions of falling bodies, as also by momemtum, the latter constantly increasing due to the gravity of the ball, it would by operation of these causes be forced to *cross* its proper vertical—as in the instances cited —and, as a consequence, it would fall to the earth "a little distance *east* of the tower."

SECTION VI.—THE ROTATION OF THE EARTH INFERRED FROM ANALOGY.

M. Biot, in Sec. 115, says :

"There are striking analogies which confirm us in this

"hypothesis. Among the planets there are several, as "Jupiter, Saturn and Mars, whose oval discs afford une- "quivocal evidence of their spheroidical figure. Observing "them with care, we discover upon the surface of their "discs, *spots* whose regular change of place and periodical ' return, leave no doubt of the rotary motion of the planet "about its shorter diameter. Why, then, should not the "earth, which resembles the planets in its spheroidal fig- "ure, have like them a rotary motion about its axis ?"

The rotations of the sun, also of Jupiter and Saturn, are inferred from the disappearance and re-appearance of spots on their discs.

As the surfaces of each of these bodies present similar phenomena of currents, whirlpools, and dashing waves of fire. the spots observed on the surface of each of these bodies have, probably, a common, although unknown, cause ; hence, whatever inference of rotation may be deduced from spots observed on the sun, will also apply to these planets.

Father Secchi, of Rome, and M. Faye, of Paris, are among recent critical observers of these phenomena of the sun.

M. Faye, after speaking of the fiery currents, whirlpools, fiery billows dashing against each other, and cyclones of fire, the *whole* visible surface of the sun fluid, and in constant motion, says:

"The whirlpools of the sun are like those of the earth, "of all dimensions, from scarcely visible pores to the enor- "mous spots which are seen from time to time."

The observed spots then are whirlpools. Father Secchi, says :

"The vaporous mass fallen back on the phostosphere "soon becomes incandescent, reheated and dissolved, and "the spot rapidly disappears."

This disappearance of the spot takes place in the photosphere of its own proper motion. In speaking of the spots, he says :

"They invade the photosphere from each part of the "*circumference.*" Again he says :

"The spots arrived at a certain high latitude, cease to "appear, but after sometime re-appear at lower latitudes, "and afterwards go on anew."

He also observed that the appearance and disappearance of spots were irregular in point of time. From these phenomena do astronomers deduce the conclusion that the *real* and *irregular* motions of the spots amount to no evidence of the rotation of the sun ? No, but conclude from these phenomena that the sun rotates irregularly, spasmodically, intermittent, so to speak, different at different times.

These observations made by these eminent astronomers, justify the conclusion that the carrier of the spots consists of flowing fiery matter forming whirlpools, which present the appearance of spots ; of their motions in *all* directions ending finally in their extinction *in* the photosphere, and that the asserted rotation of the sun has nothing to do with their appearance or disappearance, hence that these phenomena of the spots furnish *no* evidence of the rotation of that luminary.

M. Biot, in the last quotation made from him, says, in speaking of Jupiter and Saturn :

"We discover upon the surface of their discs *spots* whose "whose regular change of place and periodical return "leave no doubt of the rotary motion of the planet."

The telescope shows the surfaces of Jupiter and Saturn to be in the same fiery fluid condition, with spots, or whirlpools, moving and changing in form and appearance, and the disappearance and re-appearance of spots by their own proper motions, periodically, constitute the *whole* of the evidence of the rotation of these planets. Considering the identity of these phenomena, with those observed by

Father Secchi and M. Faye on the sun, is it possible that that the proper motions of these spots observed on the discs of Jupiter and Saturn can be *any* evidence of the rotations of these planets?

Sir William Herschell inferred the rotation of Saturn from the disappearance of a spot observable on its disc. But such rotation was regarded as finally settled by the following observation, namely:

On the 7th of December, A. D. 1876, Prof. Hall saw, through the great Washington telescope, a brilliant white spot near Saturn's equator, supposed to have been caused by an eruption. The spot spread into a streak. It crossed the planet's disc. How? By its own proper motion? No, but in consequence of the rotation of the planet, and thus rotation was inferred.

The first observation showed a spot. It then changed to a streak. The streak then spread. It will be admitted that these motions were the proper motions of the matter originally in the form of a spot.

It will be admitted that the matter observed had a proper motion as it changed from a spot to a streak, and then spread; why not then attribute the disappearance of this moving and changing matter to the same agency by which it appeared in these various changes, namely, to the proper motion of that localized matter constituting (1) the spot, (2) the streak, (3) the extended streak, rather than to the rotation of the planet, of which there is no other evidence than the proper motion of this erupted matter.

In regard to the spheroidal form of Jupiter and Saturn, this probably is the effect of the sun's attraction on the fluid mass of the surface of these bodies, as it is competent to have produced such form of the earth, when in a similar fluid state, as argued in Sec. 3.

There is nothing in the position of these bodies to prevent the sun's attraction, on the equatorial regions, from bulging these planets there, as the axis of Jupiter is perpendicular to the plane of the ecliptic, and the angle of the axis of Saturn to such plane, does not differ much from the angle of the axis of the earth to the same plane.

Mars is said to rotate in about twenty-four hours, and such rotation is inferred from the disappearance and re-appearance, periodically, of markings observed on the body of this planet.

The velocity of Mars in receding from the sun and in moving towards it, is 37,845 miles per day, or during each rotation of the planet.

As it recedes, markings on the planet would disappear each day, as the effect of increased distance, and the constantly changing angle of vision of the observer would bring into view each day markings not observed the previous day.

On the planet's approach, markings would be observed each day, as an effect of diminished distance of the planet, and, in such case, what, under the theory of rotation, would be regarded as the appearance and disappearance, periodically of the *same* markings would really be *similar* markings.

Again, as Mars, when observed through the telescope, has more the appearance of the earth than is observed of any other planet, as it shows dark and light portions of its surface, corresponding, probably, in appearance to that which our globe, with its continents, its oceans, and islands, would present to an observer on Mars, there does not appear any incongruity in carrying the analogy further by supposing the existence of a tide there, whose daily ebb and flow might account for the appearances from which the rotation of the planet is inferred by astronomers. The tide by its flow causing the disappearance of markings by their submergance, and the daily ebb, their re-appearance.

From the foregoing it may be seen that the disappearance and periodical re-appearance of the markings on

9

Mars may be—not entirely fancifully—accounted for, independent of the theory of the rotation of that planet.

As to the form of this planet being that of a spheroid, as is asserted by astronomers, the evidence seems defective, for when observed, either in conjunction or in opposition, its disc appears as a circle. In intermediate positions, it appears as an oval, which would be, when at a certain angle with reference to the observer, an effect of the action of light, whose radiation often elongates bodies. This manifestation would be capable of giving to the planet an oval form, and again a changing oblate form, at times progressive in its dimensions, and again diminishing, just as observation of Mars shows to be the fact.

The differently appearing forms of Mars are the effects of light, hence its real form is not so well established as those of Jupiter and Saturn.

When it is considered that sometimes it appears as a circle, at others, an oval; at times the oblate form is apparent, again it is not, and this form again changing, being sometimes more, then less, it is pertinent to ask, what is the real form of Mars?

Is it that of a circle, an oval or a spheroid? for its motion presents all these forms to the observer during the period of its revolution.

In reviewing the analogies relied on in proof of the rotation of the earth, we find that only three planets are mentioned as showing evidence of rotation, namely, Jupiter, Saturn and Mars. The evidence of the rotation of these planets has, in the present section, been shown to result from causes independent of their rotation.

There are 7 observed primary planets; they have been scrutinized by instruments so perfect, and the observations of them have been so minute, that shadows are observed on Jupiter on the eclipse of its satellites. Streaks and spots appear on Saturn, stars become visible beyond Saturn, and between the body of the planet and its ring, and the most delicate differences in the colors of the vapors of the sun observed.

Yet there is no evidence of the rotation of any of these bodies but the three named, except Venus, as to which it is *now* conceded by astronomers that there is no sufficient evidence of its rotation.

In view of these facts, where should we place the analogies ; in what direction do they point ?

There is no evidence of the rotation of four of the seven planets, while the rotation of the three named is inferred from phenomena which may be rationally explained as being the result of the operation of causes unconnected with the rotation of these bodies.

Is not the absence of evidence of the rotations of *four* of the seven observed primary planets more potent by analogy, that the earth does *not* rotate, than the conclusion deduced from analogy that it does, based on the supposed rotation of *three* of these bodies ?

Whatever may be the cause of the rotation of a heavenly body—did any of them rotate, a motion so contrary to the laws of motion—analogy would infer it to be common to all heavenly bodies, yet of more than one-half of the whole number of observed primary planets there is no evidence of their rotation.

SECTION VII.—THE ROTATION OF THE EARTH IN-
FERRED ARGUMENTATIVELY.

M. Biot, in Sec. 498, says :

"The earth is a globe of less than 4000 miles radius, "the sun is incomparably larger. If the centre of this lu-"minary coincided with that of the earth, its radius would "reach to twice the distance of the moon, and its volume "would fill a space of this vast extent.

"Is it not then infinitely more simple to suppose, in our
"globe, a motion of rotation already indicated by so
"many striking analogies, than to consider the immense
"mass of the sun as describing each day about us, a circuit
"of nearly 600 millions of miles ?"

It will be observed from this quotation that the rotation
of the earth is not based on the requirement of a cosmic
necessity, but is inferred from analogy, and because—for
the reasons given—"it is more simple to suppose" that the
earth rotates than that the sun revolves about it.

Most of the analogies referred to in the quotation have
just been considered.

From M. Biot's standpoint, as to the distance and magnitude of the sun, his argument is the strongest for the
rotation of the earth of any presented by astronomers, but
an answer to it may be found in the third chapter, where
it is demonstrated that the sun is but 1,000,000 miles
from the earth, having a circuit around the earth of about
6,000,000 miles. Thus—to use a military expression—
reducing M. Biot's attacking force from 600 strong to 6.

This distance of the sun from the earth, as shown in
chapter third, approximates four times the distance of the
moon from the earth.

It is known that the moon revolves about the earth, and
considering the short distance the sun is from the earth
in comparison with the distance it is estimated to be by
astronomers, it would appear nearly as "simple to suppose"
the revolution of the sun about the earth, as that of the
moon.

As analogy is frequently invoked in support of the solar
system, it may be here invoked in support of the conclusion that the sun revolves about the earth, as the moon
appears to revolve about it, and is known to do so, and as
the sun appears to revolve about the earth, analogy would
induce the conclusion that it does so.

Section VIII.—The Friction of the Tidal Wave.

In the preceding sections of this chapter, the evidence of the rotation of the earth, set forth by astronomers, has been discussed. In the present section the evidence they furnish, unintentionally, *against* such rotation will be considered.

It is affirmed by astronomers, from calculations based on the laws governing antagonizing forces, that the *friction* of the tidal wave is not only sufficient to retard the rotation of the earth, but also—which is more to our present purpose—that it actually does retard such rotation.

Prof. Newcomb, in his "Popular Astronomy," page 98, (1st Ed.) in attempting to account for the apparent acceleration of the moon's motion, does so by attributing this appearance to the *real* retardation of the rotation of the earth. He says :

"The friction of the tidal wave must constantly retard "the diurnal motion of the earth on its axis, though it is "impossible to say how much this retardation may amount "to. The consequence would be that the *day* would grad-"ually but unceasingly increase in length, and our count "of time depending on the day would be always getting "too slow."

When the calculations which have be made are considered, as to the velocity of the earth, its mass and inertia, requiring a counteracting force to overcome, contrasted with the mass, extent and inertia of the tidal wave acting on the axial velocity of the earth, as a drag-lock or break —so to speak—it would strike the mind at once, and without any calculation, that the latter would act with great power as a retarding force, which conclusion is confirmed by the calculations of astronomers, as may be inferred from the last preceding quotation from one of the most eminent of modern astronomers.

It is said by him that "it is impossible to say how much this retardation may amount to," yet on the hypothesis that the retardation of the rotation of the earth is the real cause of the apparent acceleration of the moon's motion, when it is considered that this apparent acceleration is computable, why is there not *some* retardation of the earth's rotation observed, or its extent or amount susceptible of calculation ?

Why does not the retardation of the rotation of the earth correspond with the apparent acceleration of the moon's motion, in accordance with a law of proportion ?

But, as astronomers are in possession of the facts necessary to warrant the conclusion they arrive at, namely, that the friction of the tidal wave *must* retard the rotation of of the earth, the fact that it is impossible to say how much such retardation may amount to, although it appears incredible if rotation really exists, yet it becomes of secondary importance when we are told, as we are in the last quotation, that "the friction of the tidal wave must constantly retard the diurnal motion of the earth," and that the consequences must be "that the day would gradually but unceasingly increase in length, and that our count of time depending on the day would be always getting too slow."

In the foregoing statements and admissions we have presented for our consideration the anomaly of a force which "constantly retards the diurnal motion of the earth on its axis," yet while this retarding force has been acting daily through the long lapse of innumerable centuries, it is found that it has not perceptibly, or in any measurable degree retarded such rotation even to this day, for Prof. Newcomb says, "it is impossible to say how much this retardation may amount to," * which appears as a startling

---

* A deeper philosopher says it amounts to a second in a 1000 years. This would retard rotation as much as *one* minute in 60,000 years. From what *reliable* data was this calculation made? We may conclude that "it is impossible to say."

and illogical conclusion from such premise, for this admission may be regarded as an indirect and delicate way of saying that it is not perceived to amount to anything.

Notwithstanding this constant, and, we might almost say, eternal retardation there is *no* evidence that the day was any shorter in the most distant past, while increased length of the day would be the *necessary* effect of the retardation of the velocity of the earth's rotation.

The foregoing quotations justify the following syllogism, two of whose terms, might be credited to Prof. Newcomb, namely:

1. "The friction of the tidal wave must constantly retard the diurnal motion of the earth on its axis."

2. But were the earth's diurnal motion retarded "the day would gradually but unceasingly increase in length," and it does not perceptibly or measurably so increase.

3. Then these facts constitute strong evidence that the earth does not rotate.

No astronomer will deny the first and second terms of the above syllogism. Can the conclusion, deduced in the third term be denied logically. Aside from the syllogism, it may be plainly said that if it be conceded—as it is—that the friction of the tidal wave would retard rotation did the earth rotate, and there is no such perceptible or measurable retardation, this amounts to evidence *per se* that the earth does not rotate.

Again, in any similar case, it may be logically asserted that if any force be competent *per se* to retard the velocity of a given body, and it acts on such body, and retardation can neither be perceived nor measured, these facts amount to strong *prima facie* evidence that the body said to move, really has no motion.

Section IX.—Conclusion.

In concluding this chapter, it may be truthfully said that every astronomical observation of the heavens may be as reliably made, and every calculation deduced from such observation—including those as to nutation and the precession of the equinoxes—may be found correct on the hypothesis that the earth does not rotate, as on the opposite theory.

It is worthy of observation, that not a single mathematical demonstration has been invoked in proof of the rotation of the earth, and *but one*—in another part of M. Biot's work—supposed to be in proof of the progressive motion of the earth in the orbit.

What constitutes the paramount element of weakness in the *theory* of the rotation of the earth is that every fact and every phenomenon on which the theory is based, or by which it is attempted to be supported, may be rationally inferred to result from the operations of natural laws having no necessary connection whatever with the theory of the rotation of the earth, and *independent of all hypotheses*.

If the contents of the present chapter warrant this latter assertion, it makes the theory of the immobility of the earth more philosophical, according to the standard laid down by Sir Isaac Newton, who tells us that "The main "business of natural philosophy is to argue from phenom- "ena without feigning *hypotheses*, and to deduce causes "from effects."

We will take the liberty of saying, as a negative argument against the theory of the rotation of the earth, that as the sun has as much the appearance of motion as the moon has, and as the earth appears to be without motion, it will be conceded that the burden of proof that it is not as it appears, that is to say, that it rotates, rests with those who assert it.

The proof offered by astronomers of such rotation points in opposite directions, as is shown in the foregoing sections. Is this sufficient to prove motion against the appearance of immobility ?

In "Montaigne's Essay," we find the following in Chapter xii :

"The heavens and the stars have been 3000 years in "motion ; all the world were of this belief till Cleanthus' "the Samian, or, according to Theophrastus, Nicetas of Sy-"racuse took it into his head to maintain that it was the "*Earth* that moved, turning about its axis, by the oblique "circle of the zodiac. And Copernicus has in our times so "grounded this doctrine that it may regularly serve to all "astrological consequences."

This idea of Cleanthus, or Nicetas, contains the germ of the theory of Copernicus, evolved by him nearly 2000 years afterwards.

As a consequence of this progressive motion of the earth —the sun being immovable—the earth must rotate in order to produce the phenomena of day and night.

The facts and phenomena discussed in this chapter and controverted therein by the present writer, are relied on by astronomers as evidence of such rotation.

But what is regarded, perhaps, as the firmest support, the most rational argument sustaining the theory of such rotation, is the following—presented by Copernicus— quoted from Prof. Newcomb's "Popular Astronomy":

Copernicus "explains how an apparent motion may re-"sult from the real motion of the person seeing, as well as "from the motion of the object seen, and thus shows that "the diurnal motion may be accounted for just as well "by a revolution of the earth as by one of the heavens.

"To sailors on a ship sailing on a smooth sea, the ship "and everything in it, seems to be at rest, and the shore "to be in motion.

"Which, then, is more likely to be in motion, the earth
"or the whole universe outside of it ?   In whatever pro-
"portion the heavens are greater than the earth, in the
"same proportion must their motions be more rapid to
"carry them around in 24 hours.

"Ptolemy himself shows that the heavens were so im-
"mense that the earth was but a point in comparison, and
"for anything that is known, they may extend into in-
"finity."

"Then we should require an infinite velocity of revo-
"lution.   Therefore, it is more likely that this compara-
"tive point turns, and that the universe is fixed, than the
"the reverse."

In addition to the foregoing ship and shore illustration,
as to the deceptive appearances resulting from motion, we
may suppose three parallel railway tracks quite close to
each other.

Say, for example, that we are in a car which is at rest
on the middle track.   On looking at a car on the adjacent
track, say to our right, we sometimes experience a noise-
less gliding of our car.

This apparent motion we soon find to be caused by the
real motion of the train on such adjacent track.

During the continuance of this apparent motion of our
car, if we observe a motionless train on the track to our
left, it also has the appearance, for a short time, of a motion in the same direction that our car appears to be
moving.

The apparent motion of the train to our left, results from
the persistance of visual impressions, explained in Chapter I.   Here are the apparent motions of two trains,
which are really at rest, caused by the motion of a third
train which appears to be at rest.

Again, when our train is really in motion with a great
velocity, trees, houses, fences and the landscape near the

track appear in rapid motion in a direction opposite to that of our motion.

Again, if we shut out of view all objects exterior to our car, and direct our observation exclusively to the interior of the car in rapid motion, we seem to be at rest.

If we observe from our motionless train, another train really in motion on the opposite shore line of a wide river, while it does not appear, while looking at it, to be in motion, yet after the lapse of a short time, we observe it in a new position with reference to us, which proves that it has moved.

If our train be in motion and that on the opposite shore be at rest, the apparent change of position of the latter is the same as in the former instance. Hence mere observation will not inform us which train is in motion, in the instances just cited.

Some of the foregoing illustrations show that an apparent motion may not be real, while others show that real motions may not be apparent. While the case of the two trains, on opposite sides of a wide river, shows that where one of the two is at rest and the other in motion, mere observation will not determine which one of them is in motion.

From the operation of like causes, it cannot be determined whether the observed phenomena of the heavenly bodies appearing to revolve about the earth, is a real motion, or that the rotation of the earth is the cause of such appearance.

Were the impossibility of the existence of any fact or phenomenon clearly proved, the non-existence of such fact or phenomenon would be regarded as equivalent to conclusive evidence of the non-existence of such fact or phenomenon.

While, on the other hand, the bare possibility of the existence of such fact or phenomenon, would amount *per se* to *no* evidence whatever of the existence of such fact or phenomenon.

Hence, the possibility that the earth may rotate, because did it do so, the phenomena observed would be the same did the heavenly bodies revolve about it, amounts to *no* evidence *per se* that the earth rotates.

Yet the bare possibility of such rotation is *virtually* regarded by astronomers as equivalent to some evidence of such rotation.

While all the phenomena—that cited by Copernicus and those given in connection with railways—would argumentatively show the possibility of the rotation of the earth, consistently with the observed phenomena, yet these do not furnish the slightest evidence that the earth does rotate, but are consistent with the theory, herein maintained, that the heavenly bodies revolve about the earth.

- In regard to these respective motions, it is asked in the last quotation, "Which, then, is more likely to be in mo-"tion, the earth or the whole universe outside of it ?"

In reply to this, it may be said that as the planets and the moon are conceded to be in motion it may rationally be conceded that motion of all the heavenly bodies is possible, and all of them that are visible have the appearance of motion as have the planets and moon.

The quotation continues :

"In whatever proportion the heavens are greater than "the earth, in the same proportion must their motions be "more rapid to carry them around in twenty four hours."

Reply,

On the basis of the sun's distance, as shown in Chapter II, the rapidity of its motion about the earth is not necessarily an argument against the possibility, nor even the probability of such motion, for the following reasons.

The sun, by aid of the telescope appears as a globe of fire, and in a very agitated condition.

A meteor is ignited as a consequence of its rapid motion

through a resisting medium, above the possible limits of the atmosphere, in accordance "with the mechanical theory of heat," all of which is shown in Chapter III.

Considering these facts, in regard to meteors, in connec-tion with the reduced distance of the sun as shown in Chapter II, it may not be unphilosophical to conclude that the apparent motion of the sun about the earth in 24 hours, is a real motion, and that it is ignited by its rapid motion—namely, 250,000 miles an hour—as is the meteor by its passage through the same resisting medium as that which fired the meteor, namely, that medium which is above the possible limit of the atmosphere, which is proved to exist by the fact that meteors are ignited by passing through such resisting medium beyond such limit, as is shown in Chapter III. Such ignition, in both cases, re-sulting from "the mechanical theory of Heat."

In regard to the rapid motion of the planets said to be necessary to carry them about the earth in 24 hours, it may be said that the distance of the sun is adopted by as-tronomers as a measuring rod, or unit of measurment by which the distances of the planets must be determined.

Then their distances must be reduced in proportion to the reduced distance of the sun, as found by the new measurement of such distance.

On this basis, the distance of the planet Mercury from the sun is 387,000 miles. Its velocity on this estimate is 96,000 miles an hour. Its velocity according to the present astronomy is 106,000 miles an hour. The other planets would require a greater velocity to carry them about the earth in 24 hours, in proportion to their greater distance from the sun.

The quotation continues. "Ptolemy himself shows that "the heavens are so immense, that the earth was but a "point in comparison, and for anything that is known, "they may extend into infinity. Then we should require "an infinite velocity of revolution. Therefore it is more "likely that this comparative point turns, and that the "universe is fixed, than the reverse."

Reply.

A unit, or single production does not appear to be the method of nature anywhere, but rather that of multiplication and diversity, hence, "for anything known" multiplied millions of *systems* of worlds "may extend into infinity," each having a centre of motion unconnected with ours.

What we are most concerned in here is not the conjectural or the invisible and infinite heavens, but with the finite and visible heavens.   The latter consists of the sun, moon, planets and visible stars.

As the real distances of all these bodies—except, perhaps, the moon—are unknown, as shown in Chapters I and VI, the argument, founded on the distances of these bodies, against their revolutions about the earth is deduced from a false premise, namely, the proportional, but not real distances of these bodies.

Considering that the moon and planets are not fixed, and none of the heavenly bodies have the appearance of being fixed, it does not seem "therefore *more* likely that the earth turns and that the universe is fixed than the reverse."

CHAPTER V.

DOES THE EARTH PROGRESS IN AN ORBIT?

According to the Copernican theory of astronomy, the earth progresses about the sun, moving in an orbit 570 millions of miles in circumference. Hence, the diameter of this orbit is about 190 millions of miles.

Viewed from any part of the earth's surface, the North pole of the earth appears to point on the pole star.

In order to account for the seasons, in accordance with this theory, the earth's axis is placed obliquely to the orbit and parallel to itself in every part of the orbit as the earth progresses in it.

Notwithstanding this obliquity of the earth's axis to the plane of the ecliptic, the North pole appears to point on the North star *all* the time, just as it would did the earth *not* progress in the orbit.

It is reasonable to suppose that it does so point at some time. If it really does so, at *any* position, it must diverge at least 95 millions of miles at any other position which may be as much as the one-fourth of the circumference further in the orbit, than the position first mentioned.

For example, suppose the North pole *really* to point, in December, on the North star, considering that the earth's axis is always parallel to itself; in March the axis would point away from the star 95 millions of miles, namely, the semi-diameter of the orbit.

Did it really point in September on the star, it would, in March, diverge from the star 190 millions of miles, namely, the diameter of the orbit.*

---

* By consulting an astronomical chart, showing the positions of the earth in the different parts of the orbit, in its progression about the sun, the truth of these assertions will be apparent.

This apparent constant pointing of the North pole of the earth on the North star, in all parts of the orbit, consistent with a real divergence from it, is explained by astronomers on the theory that the pole star is at such a great distance from the earth that the real divergence of the pole from the star is not apparent.

They estimate the distance of the star from the earth to be two million radii of the earth's orbit, namely, two million times the semi-diameter of the earth's orbit.

As an answer to this estimate, it has been shown in Chapter I, in a quotation from Sir John Herschell, that *no* parallax of any observed star is apparent.

In the same chapter, in a quotation from M. Biot, he says that the parallax of no *planet* even, is *certainly* known.

The estimated distance of the planet nearest the earth, namely, Venus, is about 30 millions of miles.

Calculation will show, based on these estimates, that the pole star is more than six million times further from the earth than is Venus.

Placing the pole star at this distance is expected to account for the divergence—necessary to the Copernican theory—of the pole from the star, in order to explain the apparent pointing consistent with a real divergence.

In view of the admitted fact that the distance of no star is known, can it be said that explaining the anomoly of an apparent pointing of the pole on the star, consistent with a real divergence, as an effect of the great distance of such star, is either philosophic in its method, or reliable as a deduction ?

There is an observed fact which comes pertinently to the mind, in connection with the theory of the earth's progressive motion.

For example, suppose an observer to be on the tropic of Cancer, on the 21st of June, and he observes a star on that day at or near the zenith of the observer.

On the 21st of December, after the earth is supposed to have crossed its orbit, and traveled a distance equal to 190 millions of miles, if measured in a straight line, namely, the diameter of the orbit, there the observer looks for the star he had six months before, observed in June, at the zenith in Cancer, and he finds this same star at the zenith in Capricorn, yet it is not denied that the stars are fixed.

This would be the result did the earth have *no* progressive motion.

When it is considered that the distances of the stars are unknown, hence that it cannot be known that this observed phenomenon is an effect of such unknown distance, this at first view appears at war with the theory of the progressive motion of the earth in the orbit.

It was this fact which prevented the celebrated Danish astronomer, Tycho Brahe, from accepting the Copernican theory.

As to the divergence of the north pole from the north star, it may be said—and an examination of the theory, by aid of charts will prove—that such real divergence is essential to account for the seasons, according to the theory.

Did this necessary real divergence take place—namely, not less than 95 millions of miles—would not some divergence be apparent? Even assuming it to be true that the star is distant from the earth two million radii of the earth's orbit.

In the direction of the possibility of this, it may be said, that according to astronomical estimate the sun is about 93 millions of miles from the earth.

Its extreme divergence, so to speak, measured around the curve of the earth, and extending from one tropic to the other is about three thousand three hundred miles, moving in this distance across 47° of latitude. Yet were

10

straight lines drawn from the sun to each tropic, the angle would be as perceptible as would be that formed by the minute hand of a clock in moving along its dial from 12 to 3. The change of place of the sun, marked on the earth, is observable from day to day by the eye, as small as such change is.

When it is considered on the one hand, that the estimated distance of the sun is 93 millions of miles, and that its apparent change of place, included with the limits of 3200 miles measured around the curve of the earth, is plainly observable even if measured from day to day.

And when, on the other hand, it is considered that the estimated distance of the pole star, is two million radii of the earth's orbit, and that the possible displacement of the pole from the star is the diameter of the earth's orbit, namely, 190 millions of miles, in accordance with a law of proportion there should be, in six months, a very perceptible displacement of the north pole from the star.

When the sun's shadow is visibly displaced in *a day*, when the relative distances of the sun and the star, and the distances of the sun and pole move relatively, are considered, a law of proportion, will require a visible displacement of the pole from the star in *six months*.

In this section, a deduction is made that the earth does not progress in the assigned orbit.

The facts enumerated in this section were considered in the last preceding chapter, and for another purpose than that therein contemplated.

These facts, and the deductions made from them in this section, by the writer, are as follows:

1. At 10 miles above the earth, the atmosphere has lost seven-eighths of the density it has at the surface of the ocean.

2. At this distance above the earth it is so attenuated as to be incapable of sustaining the vapors of which clouds are formed nearer the earth.

3. At 45 miles above the earth it has become so reduced in density, as to be incapable of reflecting the rays of the sun. *

4. Space, at the distance of 80 miles above the earth, is "*void* of an atmosphere for all practicable purposes." —*Herschel.*

5. Meteors in an ignited condition were generally observed at about 75 miles above the earth, and disappeared at 55 miles above it.—*Newcomb.*

6. There was "positive evidence," of these being in an ignited state, as much as 100 miles above the earth.—*Nencomb.*

7. Did the earth, moving in the orbit with the velocity of 19 miles per second, meet a meteor at rest, the earth by striking it would thus generate 600,000 degrees of heat.—*Newcomb.*

8. A resisting medium occupies space, as the pressure of such medium decreases the velocity, and reduces the visible nebulosity of comets moving in it.—*Herschel.*

The following conclusions are deduced from the foregoing enumerated facts.

1. That those meteors, observed in an ignited condition at the distance of 100 miles above the earth were *not* ignited by impact with the atmosphere, as space is *void* of an atmosphere beyond *eighty* miles.

2. That meteors at the distance of 100 miles above the earth were ignited by moving rapidly in a resisting medium, existing in space above the possible limits of the atmosphere, in accordance with the "mechanical theory of heat."

---

* The foregoing, are regarded as facts in standard works, treating of these and kindred subjects.

In view of the foregoing enumerated facts and the conclusions deduced from them, we have for our consideration a gigantic meteor, namely, the earth, and its atmosphere, progressing in the orbit with the *same* velocity as that which ignited meteors, namely, 19 miles in a second, and moving with this velocity through the *same* resisting medium—above the possible limits of the atmosphere—which ignited the meteors, and which, according to the "mechanical theory of heat," would develop 600,000 degrees of heat.

Yet the earth, with this rapid motion in the orbit, moving in the same medium as that which retards and compresses comets, and, in passing through which, meteors are ignited, wholly fails to manifest to sensation or to observation one single degree of heat known to result, expected to result, or believed to result from *this* cause, at the earth, either in its atmosphere,* on its surface, or in its waters.

When these facts are weighed and considered, do they not amount to an argument adverse to the theory that the earth progresses in the orbit?

The aberration of the light from a star, observed in a telescope, is thought by astronomers to demonstrate the progressive motion of the earth in its orbit.

So satisfied are they of this, that M. Biot expresses the general opinion when he terms it, "an incontestable truth."

This aberration is explained in connection with Fig. 17, copied from Prof. Newcomb's "Popular Astronomy."

"In Fig. 17, let S be a star, and T O a telescope "pointed at it. Then, if the telescope is not in motion, "the ray S O T emanating from the star, and entering the

---

* What would be the effect on the *atmosphere* did it strike an ether in space dense enough to compress comets, moving in such ether with a velocity of 19 miles per second? All observable effects on the atmosphere are accounted for by the operation of well known laws independent of such supposed motion.

"centre of the object glass, will pass down near the right "hand edge of the eye-piece, and the star will appear in "the right of the field of view.

"But instead of being at rest, all our telescopes are car-"ried along with the earth in its orbit around the sun at "the rate of 19 miles a second.

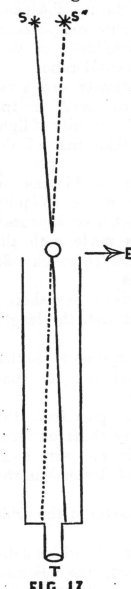

FIG. 17

"Suppose this motion to be in the "direction of the arrow; then, *while* "*the ray is passing down the telescope,* "the latter moves a short distance, so "that the ray no longer strikes the "right hand edge of the eye-piece, but "some point further to the left, as if the "star were in the direction S″ and the "ray followed the direction of the dotted "line. In order to see the star central-"ly, the eye end of the telescope must "be dropped a little behind, so that in-"stead of pointing in the direction S, "it will be really pointing in the direc-"tion S″, shown by the dotted line.

"This will then represent the appa-"rent direction of the star, which will "seem displaced in the direction in "which the earth is moving."

It will be seen from the foregoing quotation, that the phenomenon there-in mentioned occurs "while the ray is passing down the telescope," and that the cause of the observed aberration of the ray is attributed to the motion the telescope has—in common with the earth in its orbit—during the time the ray is on its travels—not from the ob-served star—but from the top of the telescope to the bottom of it.

This is the phenomenon relied on by astronomers as the "incontestible truth," which proves the progressive motion of the earth in its orbit.

In order to show that the motion of the light from the observed star, during the time it is passing down the telescope, combined with that of the earth during the same time, cannot be the known *cause* of the observed phenomenon, the following calculation of the possibility of this, and conclusion from it is submitted for consideration.

As light—according to the general estimate—has a velocity of 200,000 miles per second, and the earth in the orbit a velocity of 19 miles per second, the velocity of light then is more than 10,000 times greater than that of the earth.

In order to test the possibility of the combination of these two motions being the known cause of the obliquity of the ray as observed in the telescope, let it be supposed, for example, that observation of a star is made, with the result mentioned in connection with Fig. 17, through a 26 feet telescope, having a 2 feet object glass.

As light travels 200,000 miles in a second, how short a time would be required for it to travel 26 feet, the length of the telescope?

Calculation will prove that the time required to do this is the $\frac{1}{36,784,615}$ part of *one* second. Can such time be conceived? Can such velocity be observable?

The ray is observed in the telescope to pass from the right to the left hand of the eye piece, say about an inch. How long a time would be required to do this, were such motion of the ray caused by the motion of the earth in the orbit alone?

Calculation will show that the earth moves in the orbit an inch in the $\frac{1}{1,203,840}$ part of *one* second.

The two necessary sides of a parallelogram by which a diagonal could be formed, which should measure the obliquity of the ray emanating from the observed star, would be the horizontal line of the earth's motion in the orbit, and the direction of the ray perpendicular to this. The diagonal of this parallelogram would be formed in the $\frac{1}{1,203,840}$ part of one second, had the ray from the observed star, nearly the same velocity as that of the earth, but when it is considered that the ray travels along one side of this parallelogram, to the bottom of the teloscope in the $\frac{1}{36,784,615}$ part of *one* second, how could the diagonal of *this* parallelogram be constructed according to the law of the "composition and resolution of forces." How could this diagonal or obliquity, resulting from these two motions be, possibly, *observable*? Could such infinitessimal deviation of the ray from a straight line be measurable?

Such aberration of the ray would be invisible to the eye even with the aid of the most powerful microscope ever employed to magnify the minute, even were it capable of being used in this instance. Hence it is rational to conclude that *such* aberration has never been *seen*.

It would not be an exaggeration, were it said that the inconceivable velocity of light in moving the length of a telescope, in comparison with the velocity of the telescope itself, moving with the earth in the orbit, namely, 10,000 times *less* than the speed of light would be as the velocity of lightning to the gait of a Tortoise.

In order to conceive of the speed of light, in comparison with the velocity of the earth in the orbit, some relative comparison may be realized, when it is considered that in the time required for the earth to jog along 19 miles in the path of the orbit, light would travel far enough to cross the Atlantic Ocean 80 times, were the ocean supposed to be 2500 miles wide.

From the speed of light let it be determined how far the motion of the earth moving at a right angle, to the direction of the motion of the light coming from the star, and moving 10,000 times slower than light, would deflect the latter, and especially *visibly* deflect it, during the time it moved 26 feet in descending the telescope.

Instead, however, of the phenomenon just mentioned being evidence of the progressive motion of the earth, it may possibly result from one or more of the following enumerated facts, which are found by the observations of astronomers. It is stated by them that:

"No lens will bring all the rays of light to absolutely "the same focus.

"When light passes through a prism, its various colors "are refracted unequally, red being refracted the least, and "violet the most.

"It is the same when light is refracted by a lens, and the "consequence is the red rays will be brought to the farth-"est focus, and the violet to the nearest, while the interme-"diate colors will be scattered between."

Here it will be noticed that the refraction, namely, "change in the direction" of the ray, is not supposed to be caused by the progressive motion of the earth, yet the ray would pass down the telescope oblique to its parallel sides, and admitted to be caused by its lens, yet, from such phenomenon as this, the progressive motion of the earth is deduced, namely, the oblique direction of the ray in passing down the telescope. The quotations from Newcomb's Popular Astronomy are continued. He says:

"As all the light is not brought to the same focus, it is "impossible to get any accurate image of a star or other "object at which the telescope is pointed : the eye seeing "only a confused mixture of images of various colors.

"As larger and more perfect achromatic telescopes were "made, a *new* source of *aberration* was discovered, no prac-"tical method of correcting which is yet known.

"It arises from the fact that flint glass as compared with "crown glass, disperses the blue end of the spectrum more "than the red end.

"If the eye piece be pushed in so that the star is seen,
"not as a point, but as a small disc, the centre of the disc
"will be green or yellow, while the border will be reddish
"purple. But, in the immense refractors of two feet aper-
"ture or more, of which a great number have been produced
"of late years, the secondary aberration constitutes the most
"serious optical defect; and it is a defect, which arising
"from the properties of glass itself, no art can diminish."

If the star observed through the telescope shows but "a
confused mixture of images, as a consequence of the light
not being brought to the same focus on the lens, and if
secondary aberration constitutes the most serious optical
defect, constituting a *new* source of aberration, which is
said to arise from a property in glass itself, and, "which
no art can diminish," are there any appliances of science
so delicate and yet so reliable as to determine with mathematical certainty that the aberration, refraction, or
obliquity of the ray observed in the telescope does *not* result from some of the causes of error mentioned.

Or from a cause of error not mentioned, namely, that
the obliquity of the ray resulting from the progressive
motion of the earth, must, as a consequence of the velocity
of light, in comparison with the assigned velocity of the
earth in the orbit, be so great that *such* obliquity if existing at all would not be observable.

Such conclusion should not appear irrational, when it is
considered that there are here three distinct causes of aberration each of which, it is admitted, results in an obliquity
of the ray, emanating from a star, in its descent of the
telescope. This result, namely aberration, proceeding
from each of these causes, must be separated and measured, and one of them shown to be the effect of the progressive motion of the earth.

These causes of aberration are:

1. That which refracts light in the lens, as "no lens

will bring all the rays of light to absolutely the same focus."

2. That "new source of aberration arising from the properties of glass itself."

3. That attributed to the combined motions of light and of the earth in the orbit.

This latter named aberration or obliquity of the ray observed in the telescope, must be separated from the two former, measured, and clearly recognized as the diagonal of a parallelogram resulting from that light from the observed star, which is projected through the telescope in the $\frac{1}{36,784,845}$ part of one second, combined with the supposed progressive motion of the earth which moves across the eye piece of the telescope in the $\frac{1}{1,203,840}$ part of a second.

Can this latter obliquity of the ray be separated from the two former, and shown to result from the progressive motion of the earth? In experimenting with light on a prism, that eminent astronomer Arago, a co-laborer with M. Biot, found, in the language of M. Biot:

"Very different from what we should expect, that all "light, whether terrestrial or celestial. direct or reflected "undergoes exactly the same deviation in whatever direc- "tion it is emitted. * * * * * * * These consid- "erations with many others, prove that our knowledge of "the nature of light is yet very imperfect."

From these experiments the conclusion may be deduced that aberration of a ray would result, independent of motion, hence, that such would result from an observed star whether the earth were at rest or in motion, and second, that error in our conclusions as to phenemena observed in connection with light might arise from our imperfect knowledge of light.

In addition to the complications caused by the action of the lens on light in solving the problem of the earth's progressive motion, deduced from the aberration of light as observed in the telescope, the different motions given to the earth by astronomers, should be correctly estimated and deducted from that aberration supposed to be caused by the progressive motion of the earth in the orbit by those who believe in all these motions, and who also believe that the aberration of the light from a star observed in the telescope proves such progressive motion.

According to the conclusions of astronomers, the earth has five motions, each entirely independent of the other, namely :

1. The motion in the orbit. 2. On its axis. 3. Around the star sphere, in common with the solar system annually. 4. The earth and moon around a common centre of gravity. 5. Around the pole of the heavens, causing the precession of the equinoxes.

When the extreme minuteness of the aberration due to motion of the earth in the orbit, is considered in connection with the foregoing enumerated motions of the earth, the uncertainty of our knowledge of light, the aberration of light in connection with the lens, there is reason to doubt that the aberration of light observed in a telescope and emanating from a star would be any evidence of the progressive motion of the earth in the orbit even were it a fact that it had such motion.

In speaking of the apparent motions of the stars as confirming the theory that the aberration of light proves the progressive motion of the earth, M. Biot says :

"These very remarkable phenomena are necessary con- "sequences of the earth's annual motion; let us see "whether they are confirmed by observation.

"We find, in fact, that they are so, very exactly. *All* "the stars appear to describe annually in the heavens, "small ellipses, the dimensions of which are *precisely* "those which result from the preceding theory."

Namely, that these small elliptical orbits are only apparent as consequences of the annual motion of the earth as proved by the aberration of light from a star observed in the telescope.

As partly an answer to this, we quote from Prof. Newcomb's "Popular Astronomy," Part IV, where he says, in speaking of a star: "If it be moving directly towards us, "or directly away from us, we could not see any motion at "all. * * * * * * The complete motion of the "stars cannot therefore, be determined by mere telescopic "observations. * * * * * * The motion of each "individual star is so entirely different from that of its fel- "lows, as seemingly to preclude all reasonable probability "that these bodies are revolving in definite orbits around "great centres of attraction.

"If the stars were moving in any regular circular orbits, "whatever, having a common centre, we could trace some "regularity among their proper motions, but no such regu- "larity can be seen. * * * * The stars in *all* parts of "the heavens move in *all* directions, and with all sorts of "velocities. * * * From the observation of the motion of a "single star, it is *impossible to decide* how much of this ap- "parent motion is due to the motion of our system, and "how much to the real motion of the star."

If these quotations from Prof. Newcomb are true,—and in point of fact there can be no doubt of this—how can it possibly be true, as asserted by M. Biot, that: "*All* the "stars, (namely 50 million), appear to describe annually "in the heavens small ellipses, the dimensions of which are "*precisely* that which result from the preceding theory," namely, the progressive motion of the earth in the orbit.

By turning to Chapter I, a quotation from Sir Jno. Herschell may be found, wherein he says that the most minute parallax of any observed star has not been found, and that so perfect are the instrumentalities for this pur- pose, that had there been a parallax so minute as one sec- ond, it could not have escaped instant detection. One second would not be of the dimensions of one-half the thickness of the wire on the micrometer.

Then it cannot be true, as asserted by M. Biot that these small elliptical orbits observed of all the stars, are merely optical illusions *caused* by the progression of the earth in the orbit.

Because, in order to confirm the theory that the differing apparent dimensions of the orbits of the stars were due to the real progression of the earth in the orbit, the distance of the earth from the observed star would be necessary to be known in order to determine whether the apparent dimension of the orbit of the observed star were such as were due to such distance, and to the progression of the earth in its orbit, and as there is no parallax of any observed star, as stated by Herschell, such distance cannot be known.

And this distance must be known in order to determine whether the apparent dimensions of the orbit of the star, is that due to the real progression of the earth in its orbit, or not.

Besides, Prof. Newcomb virtually, although indirectly and of course unintentionally, confirms the correctness of this conclusion when he says : "From the observation of "the motion of a single star, it is impossible to decide how "much of this apparent motion is due to the motion of our "system, and how much is due to the *real* motion of the "star."

It will be noticed, in this quotation, that Prof. Newcomb does not say how much of the apparent motion of the star is due to the real motion of the earth in the orbit.

Taking the foregoing quotations from Prof. Newcomb as authoritative, as they doubtless are, it cannot be true, as asserted by M. Biot, that the elliptical orbits observed of the stars are merely apparent, due to the real progression of the earth in the orbit. Hence, that the dimensions of such stellar orbits do not prove the progression of the earth in its orbit.

Another argument of M. Biot may properly be here noticed, wherein he attempts to prove the progressive motion of the earth in the orbit by the use of an analogy, comparable to the aberration of light. He says:

"Thus, if a person in a steady boat under sail, or pro-
"pelled by steam, should endeavor to hold a tube, so as to
"catch drops of rain or snow that may be supposed to fall
"perpendicularly, he would be obliged to incline the tube
"in the direction of his motion, in order that the drops
"might describe lines parallel to the axis; that is, the tube
"must form the diagonal of a parallelogram, the sides of
"which would be the velocity of the boat, and that of the
"falling drops.

"If the earth has really a progressive motion around
"the sun, it must produce a similar effect upon the light
"coming from the heavenly bodies; and the impression
"of this light upon the eye cannot take place in the actual
"direction of the rays.

"This is what actually happens, and it constitutes what
"is called the aberration of light."

Here it will be readily perceived that the analogue consists in the moving boat representing the earth progressing in the orbit; the stationary cloud stands for the motionless star; the tube for the telescope, and the rain falling into it in a direction oblique to the normal direction of the fall of water-drops is in lieu of the ray from the star descending the telescope obliquely.

While all that is said in the quotation is true yet it does not follow that the boat must move in the one case, nor the earth in the other in order to produce the observed phenomena in the tube and in the telescope.

The force of conviction resulting from a mathematical demonstration consists in the fact that no reasonable hypothesis is possible, adverse to the demonstration.

And where such hypothesis is possible that which is claimed as a mathematical deduction can, in no sense be considered a demonstration.

The fallacy of the deduction made in the last preceding section, results from first assuming the progressive motion of the earth to be true, and then in attempting to prove such motion by observed phenomena which are possible on another and opposite theory.

In order to illustrate the truth of this assertion let it be supposed, for example, that the boat mentioned in the last quotation were really without motion, and a cloud was rapidly passing directly over it, and from which rain were falling perpendicularly ; were the tube mentioned held diagonally to the direction such rain were falling and in a direction opposite to that in which such cloud were moving, it will not be doubted that it would be possible for the rain to descend the tube oblique to the direction the cloud were moving and to that in which the rain were falling from the cloud.

Suppose it were not known which was in motion, namely the boat or the cloud, would the obliquity in the direction of the fall of the rain *necessarily* prove that the boat, and not the cloud, had motion?

Again, suppose it were not otherwise known—and it is not—that the earth has a progressive motion, would the obliquity in the direction of a ray from a star, as observed in the telescope prove such motion any more than it would prove the motion of the observed star?

It is difficult to keep some stars in view in the telescope —Sirius, for example—in consequence of their apparent rapid motion.

All the stars have apparent motions in small orbits, and in addition all observed seem to have a drift westward.

Were it supposed that the earth were motionless and the stars in rapid motion as they appear to be, were the telescope pointed on a star as is explained in connection with Fig. 17 the ray would appear on the right hand side of the eye piece of the telescope.

The rapid apparent motion of the star would result in the ray crossing the eye piece and appearing on the left of it, and were the telescope fixed, the star would rapidly disappear from the field of view.

Would not these phenomena as clearly prove the theory that the stars have motion, as the opposite one, namely that the earth has a progressive motion, for the phenomena in support of either theory is the same?

And this is not the former named theory as rational as the latter?

The distances and magnitudes of the stars are entirely unknown.

They all appear to have progressive motions.

There is no sensible evidence that the earth has such motion.

All the observed heavenly bodies have the appearance of having proper motions.

Some of these, for example, the planets and their satellites, and the moon are known to have such motions.

In regard to the aberration of light observed in the telescope proving motion of the earth, or star, the difference of the *theory* is *all* that makes the difference in the deduction.

For the reason that the phenomenon relied on as demonstrating the progressive motion of the earth, would prove a theory of the progressive motion of the stars ; for the phenomenon relied on to prove the progression of the earth, namely, the obliquity or aberration of the ray eminating from an observed star, would be observed in the telescope did the star have a progressive motion, hence proves such motion as fully as it does the progressive motion of the earth.

The revolutions of the planet Venus and the earth in their respective orbits, together with the observed phenomena of Venus in connection with the geographic parallels of the earth constitute the basis of an argument adverse to the astronomical theory of the progressive motion of the earth in its orbit.

It is an observed fact, and is asserted in astronomical works of authority that Venus in progressing in its orbit obliquely crosses the geographic parallels of the earth.

M. Biot—Sec. 416—in speaking of the directions of the motions of the planets Mercury and Venus implies this also when he says :

"We perceive that they do not continue in the same "parallel to the equator, nor do they move *exactly* in the "plane of the ecliptic."

"But all the planets, except those lately discovered, de-"viate *very* little from the last plane."

The inclination of the orbit of Venus to the plane of the ecliptic is 3°, 23', 27", namely less than 3½°.

The earth, according to the theory, as is well known, progresses along the path of the ecliptic from west to east.

The axis of the earth is parallel to itself throughout its revolution in the ecliptic.

The obliquity of the ecliptic to the sun—as also that of the equator of the earth to the ecliptic—is about 23½°.

As a result of this theory of the earth's progression, the vertical rays of an immovable sun pass over the equator obliquely from one tropic to the other, as a consequence of the progressive motion of the earth along the oblique path of the ecliptic, while the angle of the earth itself to the plane of the ecliptic remains unchanged during its entire revolution.

11

The equator is 23½° from each tropic, and the direction of the ecliptic is oblique to the equator, and extends from one tropic to the other.

The phenomena observed of the progressive motion of Venus across the parallels, should be in harmony with the theory of the earth's progression in its orbit, in order to render such theory tenable.

Venus makes a revolution in its orbit in 224 days, the earth in 365 days.

According to the astronomical theory, Venus, like the earth progresses from West to East, and it is found that the orbit of Venus is nearly coincident in direction with that of the ecliptic, in which the earth progresses—by the theory.

Were it supposed, for example, that both Venus and the earth made their respective revolutions in the *same* time, it will be readily perceived that while the rays of the sun would appear to cross the earth's parallels, and the earth's motion would be across the celestial parallels, Venus would not cross any terrestrial parallel, nor would it have the appearance of so doing, because, having the same velocity as the earth and making a revolution in the *same* time as the earth, as is the case just supposed, Venus could neither advance nor recede from its relative position towards the earth ; hence, in such case, over whatever parallel of the earth Venus were vertical at the commencement of its revolution, such would be its position relative to that of the earth at its close.

From which, the conclusion is deducable that the apparent velocity with which Venus would cross the parallels, in moving in the same direction as that of the earth and parallel to it, would be in proportion to the *relative* velocities of Venus and the earth.

In regard to the various observed positions of Venus in different parts of its orbit, with reference to those of the earth in its orbit, as also to the sun, M. Biot, in Sec. 412, says :

"When Venus appears as an entire circle, its diameter "is very small, not amounting to more than 10″. It is then "beyond the sun with respect to the earth. On the con-"trary, when its phases diminish, and its illuminated face "is turned more and more from us, its apparent diameter "is on the increase, which indicates that it is now nearer "the earth. Lastly, in the interval between its disappear-"ing in the evening and re-appearing in the morning, we "sometimes see it moving over the sun's disc with the ap-"pearance of a round black spot, and its apparent diame-"ter is then the greatest it ever attains during a revolu-"tion, amounting sometimes to 59.8″, or very nearly 1′."

M. Biot, in Sec. 413, says :

"The orbit of Venus does not embrace the earth ; since, "if it did, this planet would sometimes be in the part of "the heavens opposite to the sun, the earth being between "them, which never happens.

"Neither is its orbit wholly beyond the sun with respect "to the earth ; for if it were, Venus would never appear "between the sun and the earth ; whereas, it is sometimes "seen passing over the sun's disc.

"Lastly, while the sun moves, or appears to move in "the ecliptic, Venus never deviates beyond its customary "limits, and its oscillations about the sun, or its *elonga-"tions*, are always nearly of the same extent.

"These facts, taken together, manifestly prove that Ve-"nus moves about the sun in an orbit, which returns into "itself, and which accompanies the sun in its apparent "elliptical motion."

From these well ascertained and acknowledged facts, the conclusion is imperative—if we add the facts that the velocities, the times of revolution and the circumferences of the orbits of Venus and the earth are different—that Venus during its revolution is sometimes moving in the same direction as the earth, sometimes moving in an opposite direction, and again at different times at various angles to the earth.

How *should* Venus appear to cross the earth's parallels —according to the theory of the earth's progression in the orbit—at a time when Venus and the earth were on opposite sides of their respective orbits, say at a time when the earth were moving eastward and Venus were moving westward ?

When in such and similar positions, owing to the immensity of each of their orbits, Venus and the earth would move along almost straight lines, for long distances before the slight curves of their orbits would place them at a perceptible angle to each other.

If this view be correct, and Venus and the earth were in the relative positions just mentioned, Venus, from the earth, would appear to cross the earth's parallels with a rapidity due to the *two* velocities, namely, that due to its own proper velocity in one direction, and that due to the earth's velocity in an opposite direction, which would be equivalent in apparent results, so far as the phenomenon of Venus crossing the parallels would be observed, as though the earth's progressive motion were instantly stopped, and Venus had, at the same instant, the previous velocity of the earth *added* to its own proper yelocity.

The appearance, in such case, would be analogous to that of two railway trains—on parallel tracks, close to each other—moving in opposite directions, say, each with the same velocity ; in this case, each could appear to the other, at the time they were passing, to be moving with double its real velocity.

And in a certain sense, each train has double the velocity properly belonging to it, namely, in the sense that the observed train actually does move past the other in one-half the time it would were the train of the observer without motion.

For example, were plainly visible, vertical parallels placed on the side of each train, each train would move past such parallels on the other train in one-half the time —supposing the velocities of the trains equal—that they would were either train at rest, and the other in motion with its own proper velocity.

Such apparently increased velocity of each moving train, observed from the other, is strikingly perceptible, by comparison, after the train just passed has gone from the view, by observing stationary objects on the roadside.

In such case, our train does not appear to move past such objects with the rapidity observed of the passing train, when moving in an opposite direction.

By analogy to the illustration of the two trains moving past each other in opposite directions, were observers on Venus and on the earth observing the opposite body, as each moved past the other in opposite directions—supposing such motions perceptible to the observers—the supposed observer located on Venus would see the earth progressing with a velocity which would not be observable in any other relative position of Venus and the earth, and in such case, the observer on the earth could see Venus cross the parallels of the earth with a velocity which would not be observable were these two bodies placed in any other position relative to each other.

Just as an observer on one of the supposed trains would observe the vertical parallels, on the side of the train nearest him, move past with a rapidity due to the velocities of both trains, which double velocity would not be observed were the train of the observer in motion while, at the same time, the train on the other track were at rest.

The difference in the times of revolution of Venus and the earth, results more from the comparative circumferences of their orbits, than from the difference of velocity.

According to astronomical estimate, the circumfereence

of the orbit of Venus is not far from 408 millions of miles. That of the earth is 570 millions of miles.

According to this estimate, the orbit of the earth is much more than one-third greater than that of Venus.

While the orbit of the earth is thus much larger than that of Venus, its velocity is less, although these velocities more nearly approximate than their orbits.

Venus, in round numbers, progresses in its orbit with a velocity of 75,000 miles an hour, and the earth with a velocity of 68,000 miles an hour.

As has been said, Venus makes a revolution in its orbit in 224 days, and the earth in 365 days.

As consequences of the difference in times of revolution of these bodies, and the differences of their velocities, there must be great and frequent differences in their relative positions to each other, and that such really is the fact, plainly appears in the last quotation made from M. Biot's work.

As a result of these different positions of Venus, relatively to those of the earth, Venus should appear to cross the parallels of the earth with observably different velocities.

In order to establish the correctness of this conclusion, suppose, for example a great many straight lines to pass through the sun's centre, and to be continued until they each crossed the orbits of Venus and the earth.

When these moving bodies were both at the same time in such position with reference to each other, as would place them on any one of these supposed lines, and were both moving parallel to each other in the *same* direction. Venus, in such ease, would appear to cross the parallels of the earth, with a greatly diminished velocity, namely, with that equal to the *difference* between the two real velocities, namely, those of Venus and the earth.

This apparent velocity, or difference between the two real velocities would be equivalent to 7000 miles an hour for Venus, as this difference may be deduced from the relative velocities of these bodies before mentioned.

This would be the same invisible results, so far as Venus' apparent crossing the parallels of the earth would be observed, as though the earth were without progressive motion, and Venus crossed its parallels with a velocity of 7000 miles an hour.

On the other hand, were Venus and the earth on any one of the straight lines just mentioned, on opposite sides of their respective orbits, and thus moving in *opposite* directions, at and near such line the apparent velocity of Venus—the observed body—would be that due to its proper velocity, together with that of the earth's, so far as the apparent speed of Venus could be observed on the earth, crossing the parallels.

These combined velocities, namely, that of the earth's in one direction, and that of Venus' in an opposite direction would be, as to the observation of Venus crossing the parallels, practically the *same* as though the earth's progressive motion were instantly stopped and its just previous velocity were at the same instant transferred to Venus, as just stated.

The observed result of this would be, during the time these two bodies were on and near the straight line mentioned, that Venus would progress with its proper velocity, namely, 75,000 miles an hour, and would also appear to have that of the earth's, namely, 68,000 miles an hour, which real and apparent velocities would be equivalent to a velocity of Venus of 143,000 miles an hour.

This would give Venus the appearance of crossing the ɼarallels with the increased velocity due to this apparent addition, at the time the two bodies were passing each other, just the same as in the instance cited of the two railway trains passing each other in opposite directions.

This apparent velocity of Venus, when it and the earth were on opposite ends of such straight line, would diminish in proportion to the magnitude of the angle formed between the two bodies, and as it became greater, as it would, as the two bodies were removed from such first position.

With the velocity of the earth apparently added to that of Venus', during the time the two bodies were on and near the straight line mentioned, and while they were parallel to each other, or nearly so, and moving in opposite directions, Venus would have the appearance to an observer on the earth, of crossing the earth's parallels with something more than 20 times the velocity it would appear to have at another time, namely, when Venus and the earth were on one of these supposed straight lines and the two bodies were moving parallel to each other and in the *same* direction.

For in this latter supposed position, Venus would progress—as before said—with an apparent velocity of 7,000 miles an hour across the parallels, namely, with the difference between the actual velocities of Venus' and the earth's.

While in the former relative position, Venus should appear to cross the parallels with a velocity due to 143,000 miles an hour, which in comparison with an apparent velocity of 7,000 miles an hour, is more than 20 times greater.

In order to show the different positions of Venus and the earth in their orbits, with reference to each other, as effects of their different velocities and times of revolution —although somewhat exaggerated in results—let it be supposed, for the purpose of illustration, that the orbits of these bodies were squares.

Venus makes a revolution in 7½ calendar months; the earth in 12 months. Venus would make one-fourth of a revolution, moving from West to East, namely, one side of the square, in 1⅞ months. The earth would progress the length of one side of the square of its orbit, in 3 calendar months.

Suppose the two bodies to move simultaneously eastward, say, for example, from the northwest angles of their respective squares.

When Venus had arrived at the northeast angle of its first side of the square, the earth would not have arrived at the northeast angle of the first side of its square, in consequence of its slower velocity and larger square, and would have traveled in 1⅛ months but little more than two-thirds the length of such side.

The two bodies moving in the same direction, and parallel to each other, the earth above Venus, Venus would move with the velocity of the earth plus the difference of their respective normal velocities, namely, with a velocity of 7,000 miles an hour.

These figures measure the *excess* of the velocity of Venus over that of the earth.

When these bodies were in the relative positions resulting partly from these different velocities, and partly from the greater orbit of the earth, Venus would have turned the first angle of the square of its orbit, formed by the first and second sides of it, and would be moving along such second side.

Venus would *then* move in the direction of a right angle to that in which the earth were then moving, for the earth would be then moving along the first side of the square of its orbit, parallel to the first side of the square of Venus' orbit, while Venus would be moving along the second side of the square of its orbit, hence, at a right angle to the direction of the motion of the earth.

The effect of these relative positions of the two bodies on the apparent velocity of Venus, in crossing the earth's parallels, would be that due to the earth's velocity *alone*.

For it will be readily conceived that did Venus move in a direction at a right angle to that of the earth's motion

—whatever might be the forms of the orbits—it could not, as an effect of its own proper motion, cross any of the earth's parallels, and whatever might be the appearance of its doing so, would be the effect of the earth's progressive motion alone.

At the terminus of 3¾ months from the time Venus had departed from the northwestern angle of its supposed orbit, it would have arrived at the southeastern angle of its square, having traveled the distance included within two sides of it, equal to one half its revolution.

When arrived at the terminus of such second side, the planet commences its motion on the third side of the square, moving westward.

At the time it commences to move in this latter direction, the earth will have turned the angle formed by the junction of the first and second sides of its supposed square orbit, and will have proceeded along such second side about one-fourth the length of such second side.

When these last named relative positions of Venus and the earth are considered, it will be manifest that the position of Venus, with reference to the parallels of the earth, will be that of an acute angle to them ; in fact, for a short time, Venus would be nearly at a right angle to their direction on the earth, and in such position Venus would be very nearly parallel with the earth's surface. That is to say, such would be the case at the commencement of Venus' progress on the third side of its supposed orbit, at which time the earth would be not far from the beginning of the second side of its square orbit.

Venus would reach the western terminus of the third side of its square, in 5⅝ months from the time it had started on its supposed revolution.

By the time Venus had arrived at the western end of the third side of its square, the earth would have begun at the east to move westerly along the third side of the square of its orbit. And as Venus commences to move along the fourth side of its square, the earth will have progressed a short distance on the third side of its square.

Again, for a short time, when the bodies were in these relative positions, Venus, as seen from the earth, would be at an acute angle to the direction of the parallels, and would be nearly parallel with the earth's surface ; hence, would appear to cross the parallels very slowly, during the apparent continuance of these relative positions, which speed in apparent crossing the parallels would appear to increase as Venus passed further along the fourth side of its square, and the earth moved further along on the third side of its square.

Now, while this supposition of square orbits, imagined for the purpose of illustration, more clearly shows the different apparent velocities with which Venus would cross the parallels of the earth as results of its different relative positions towards the earth moving in its orbit; yet these differences on this supposition, would be greater, perhaps more clearly defined, and certainly more sharply distinct, than could be observable where the orbits of these bodies elliptical, as is the form according to the theory.

The supposition of square orbits may serve to show by way of illustration, and as on approach to an approximation, the apparent and different velocities with which Venus should cross the earth's parallels to what really would be observable were these estimates made for elliptical orbits.

A calculation of the relative positions of Venus and the earth, and inferrentially of the different apparent velocities of Venus in crossing the parallels of the earth, may be found in the following table, the calculation being made on the basis that the orbits of these bodies are elliptical in form.

This table is intended to show the different positions of Venus in its orbit with reference to those of the earth in its orbit, and thus to attempt to show with what apparent velocities Venus should cross the earth's parallels on the theory of the earth's progressive motion in its orbit.

These different positions of Venus in its orbit with reference to those of the earth, are results of differences of the circumference of the orbits, times of revolution and of difference of velocity of these two bodies.

Proper proportional orbits of Venus and the earth were constructed. These orbits were intersected by lines dividing each into four equal parts, each part representing one-fourth of a revolution, and the position of each body, with reference to the other at the terminus of each one-fourth of a revolution, in accordance with the respective velocities, were noticed, and these one-fourth revolutions of Venus were made notes of. This method of determining the relative positions of the two bodies was computed for four years.

These different positions of Venus in the orbit with reference to those of the earth, were noted in order to determine, in accordance with the laws of motion, with what velocity Venus should appear to cross the parallels of the earth, in accordance with the theory of the progressive motion of the earth.

The positions of Venus and the earth with reference to each other, for each one-fourth of a revolution of Venus, computed for four years, as follows :

First Quarter, 1st year, V. and E. nearly parallel, moving same direction.
" " 2nd " " nearly opposite, moving opposite direction.
" " 3rd " " 40° from straight line, from East to South
" " 4th " " same direction, V. and E. parallel.

Second Quarter, 1st year, V. and E. angle of 40°.
" " 2nd " " nearly opposite, moving opposite direction.
" " 3rd " " moving same direction, angle 40°.
" " 4th " " " " " " 30°.

Third Quarter, 1st year, V. and E. nearly parallel to earth's surface.
" " 2nd " " moving in opposite direction, angle 20°.
" " 3rd " " moving same direction, angle 45°.
" " 4th " " " " " " 30°.

Fóurth Quarter, 1st year, V. and E. at angle 25°.
    "   "   2nd "   "   moving same direction, angle 30°.
    "   "   3rd "   "   nearly parallel, same direction.
    "   "   4th "   "   moving opposite directions, ang'e 28°.

Upon an analysis of the foregoing table it will be found that in the short period of even one year, Venus will progress in its orbit in a direction parallel to the earth's surface; again, opposite in direction to that of the earth; and sometimes in the same direction; at one time these two bodies will be moving parallel to each other, either in the same or in an opposite direction, and again will be at various angles to a straight line drawn from the earth to the sun.

According to the laws of relative motions, taken in connection with the theory of the progressive motion of the earth in its orbit, Venus should, at the earth, be observed to cross the parallels and meridians of the earth with very different apparent velocities.

The next question considered will be what are the phenomena, observed in connection with Venus in crossing the earth's parallels; and as germane to this question, how the sun crosses these parallels.

The irregularities of the sun in declination are slight, in passing from tropic to tropic, and there is an uniformity in the recurrence of the irregularities.

This may be observed on consulting the astronomical tables where the sun's declinations are recorded.

M. Biot, Sec. 160, says : "It we take the differences of "these declinations it will be seen that they are to a certain "degree regular and symmetrical on opposite sides of the "equator, but the progress of these variations are unequal. "They are more rapid when the sun is near the plane of "the equator."

On further investigation they will be found *uniformally* so.

In speaking of the longitude of the sun, M. Biot says,

Sec. 172 : "In these tables the differences of longitude
"from day to day, may be determined with great accuracy,
"as they are the same every year, and return in the same
"order."

In speaking of the periodic changes of the sun's motion
from year to year, he says, Sec. 179 : "That they com-
"pensate each other by returning through the same de-
"grees ; so that they make the sun oscillate continually
"through a mean state, from which it departs but little."

These quotations show that the apparent motions of the
sun about the earth and between the tropics, that is to say,
in both latitude and longitude are generally regular, but
at times accompanied by slight irregularities, and that
these latter are regularly compensated within the year.

The foregoing quotations are sufficient of themselves
alone, without observation, the regularity and general
uniformity with which the sun crosses the parallels of the
earth.

The next question claiming consideration is the relations
of Venus to the sun.

On this point M. Biot says, in Sec. 413 : "Venus moves
"about the sun in an orbit which returns into itself, and
which accompanies the sun in its apparent elliptical mo-
"tion."

In speaking of Venus and Mercury, Sec. 416, he says :
"These two planets, as seen from the earth, *always* appear
"to accompany the sun. Their angular distances never
"exceed certain limits, which are nearly constant. The
"same cannot be said of the other planets ; these diverge
"from the sun to all possible angular distances.

The knowledge here imparted is obtained by observation, as it is observed that Venus and the sun are always
together—so to speak.

In whatever position the sun is observed in the heavens,
whether over the tropics or equator, and at all intermediate positions in the heavens, between those where the

sun may be observed, there is Venus also observed, occupying the same relative position toward the sun, Venus must as a consequence, and which the observed phenomena proves, cross the earth's parallels with the regularity and uniformity observed by the sun.

Taking into consideration all the foregoing quotations made in this section, together with the observed phenomena, the following conclusions are reached, namely :

1. That both Venus and the sun are observed to cross the meredians and parallels of the earth in an oblique direction.

2. That the sun—as deduced from astronomical tables —crosses them with a generally uniform velocity and regularity, and when irregular in its velocity in crossing them, this is compensated on its next return thus producing a general uniformity.

3. That Venus is as constant to the sun as a satellite to its primary, being always observed at the same angular distance from it, and appears to move as the sun does about the earth, and from tropic to tropic, with the same uniform velocity that the sun appears to have.

4. That Venus appears to cross the meridians and parallels of the earth with the same apparent velocity, and uniformity that the sun does.

With a view of contrasting the observed phenomena just enumerated, with the phenomena which would be observable—and are not—lid Venus cross the parallels in accordance with the theory of the earth's progressive motion in its orbit, the following analysis and enumeration is made, namely :

1. When Venus and the earth were progressing in the same direction, and were at opposite extremities of the same straight line drawn through the centre of the sun, before mentioned, the apparent velocity of Venus, at and near such line observed from the earth, in crossing the

parallels would be that of the *difference* of the real velocities of these two bodies.

2. At the time both bodies were on opposite sides in their
orbits, and at opposite extremities of the same straight line
mentioned, and progressing in opposite directions, Venus
would appear to cross the parallels with a velocity due to
its own proper motion, combined with that of the earth's.

3. When Venus and the earth were progressing at, or
near a right angle to the directions of each others motion,
Venus would cross the parallels with a velocity due to the
motion of the earth *alone*, which apparent velocity would
be different from either of the two former.

4. At a time when Venus was at an acute angle to the
direction of the parallels, and at the same time nearest
parallel to the earth's surface, Venus would cross the parallels with less apparent velocity than would be observable were it at any other position relative to that of the
earth.

5. When Venus and the earth were progressing in the
same direction, and Venus were on a line with the axis of
the earth, Venus would not appear to cross the parallels
at all, for while in this position it and the earth would
both be progressing in a direction parallel to the parallels.

These differences in the apparent velocities of Venus,
observed from the earth, did Venus continue on or cross
the parallels, according to the theory of the earth's progressive motion, are necessary results of the following
causes, namely :

1. The difference of the velocity of Venus from that of
the earth.

2. The difference of circumference of their orbits.

3. The two bodies moving along the same plane, namely, the ecliptic.

4. Resulting in the two bodies making their respective
revolutions in different times.

What should be the phenomena observed on the earth, as results of the different apparent velocities of Venus on crossing the parallels from the causes just stated, in connection with the theory of the earth's progressive motion ?

The phenomena observable would be, that Venus, at one time, would have that apparent rapid motion across the parallels, which would result from an apparent velocity of 143,000 miles an hour, such velocity being practically the result of the apparent combination of the velocities of Venus and the earth, at the time these bodies were progressing on the opposite sides of their orbits, and moving in opposite directions, as stated.

At another time, Venus should be observed to cross the parallels with an apparent velocity twenty times *less* than that just mentioned, namely, with an apparent velocity due to the real velocity of 7,000 miles an hour. This being the difference in the relative velocities of Venus and the earth, at a time when these bodies were moving in the same direction and parallel to each other, as before stated.

At another time, Venus should appear to cross the parallels with a velocity due to that of the earth in its orbit. This would occur at the time when the direction of the progressive motions of the earth and that of Venus were at or near a right angle to each other.

This latter velocity of Venus would be apparently, greater than that last mentioned, and less than the first mentioned.

At another time, Venus should appear to cross the parallels with a velocity less than any of these before mentioned.

This would take place at a time when Venus were progressing in a direction nearly parallel to the earth's surface. At such time Venus would appear to cross the parallels very slowly, if observable at all.

At another time Venus could not be observed to cross

12

the parallels at all. This would occur at a time when Venus and the axis of the earth were on the same straight line, namely at a time when the straight line of the earth's axis—if prolonged—would pass through Venus. In such position Venus would move parallel to the earth's parallels, and the earth progress parallel to the progressive motion of Venus.

To the last foregoing views in regard to the apparent velocities of Venus in crossing the parallels, as results of its different positions, relative to the earth, it might be objected that Venus might not remain in the different relative positions a long enough time to cause their effects to be observable on the earth, yet such objection would scarcely be tenable when it is considered that Venus, in making its revolution, follows the long line of an ellipse 408 millions of miles in length, which distance it completes in 224 days, that the orbit of the earth of like form, has a circumference of 570 millions of miles. along which it travels in 365 days, while Venus on an average crosses from one parallel to another in something less than four days.

When it is considered that the distances traveled by the earth and Venus, and the long periods required to make their revolutions, and the further fact that in the short period of four days, the orbits of Venus and the earth would continue to be almost straight lines, the appearances of Venus in this short time would be observable as it crossed from one parallel, and through another with the apparent velocities due to the different relative positions it were in, with reference to the earth.

The following conclusions are deduced from the foregoing, namely :

1. Venus should cross the parallels and meridians of the earth with different velocities which should be observable as effects of the progressive motion of the earth in

- the orbit, and such are no more observable of Venus than of the sun.

2. The observed phenomena of Venus crossing the parallels and meridians are adverse to the theory of the earth's progressive motion.

3. If the two foregoing conclusions be tenable, the theory that the earth progresses in an orbit about the sun is untenable.

CHAPTER VI.

THE DISTANCES OF THE PLANETS.—THE PROGRESSIVE MOTIONS OF THE MOON AND STARS.

As the sun, by the new measurement, is but one million miles from the earth, the planets on this basis cannot have the masses and distances given them by astronomers, for had they such they could not be in that proportion to the reduced mass and distance of the sun required by the law of attraction.

This assertion will be apparent when it is considered that by the new measurement the sun's distance is diminished to one million miles, and its mass reduced to that in corresponding to that of a diameter of 9483 miles.

This reduction of the distance of the sun places it at the $\frac{1}{93}$ part of the astronomical distance; this necessarily—in accordance with a law of proportion—requires a corresponding reduction in the masses and distances of the planets.

The astronomical estimates of the distances of the planets are made both theoretically and practically.

It becomes important to consider and to attempt to show the insufficiency of the foundations on which these estimates are built.

One of the practical methods most relied on for this purpose is that by parallax. This method, for such purpose, has been considered in Chapter I.

By referring to that chapter, the reader may there find that astronomers do not place implicit and undoubting faith in its sufficiency for finding the real distances of either stars, planets or the sun, and the method so applied may be regarded as uncertain, hence unreliable, as a deduction from the facts stated in that chapter.

There is a method employed by astronomers, by which the motions of the planets are determined, which, were it reliable, would not only establish their velocities, but from which could be deduced, by a mathematical formula, their distances also. In this latter view it may be proper to analyze the sufficiency of the method *per se.*

This method may be understood from the following quotation from M. Biot's work, where he says—Sec. 151.

"In determining the motions of the heavenly bodies, we "observe the point each day in which they are found on "the celestial sphere, and afterwards we ascertain the form "of the trajectory, from the condition, that the body has "passed successively through all these positions in the "order in which they were observed."

The uncertainties of this method may be illustrated by supposing the observer of the motion of the planet to be at E. Fig, 38.

Suppose, for example, that the true trajectory of the curve were P P″ and the orbit 3, 6, 4, 5. As all circles have the same number of degrees, it may be found that P P″ is in the same proportion to the circle 3, 4, 5, 6, that 8 9 is to the circle A, B, C, D, then there are the same number of degrees in the arc P P,″ as in the arc 8 9 of the circle A, B, C, D.

To an observer at E—the earth Fig. 38.—the line of vision to the observed planet, namely, from point to point of the trajectory, would be that coincident in direction with the lines E P 8 and E P 9, and at the extremities of these lines, between which is the trajectory resulting from the planet's motion between these points.

Yet, possibly, the observed planet might be in motion in the orbit 3, 6, 4, 5, or in the orbit A B C D, and from observation it could not be determined in which.

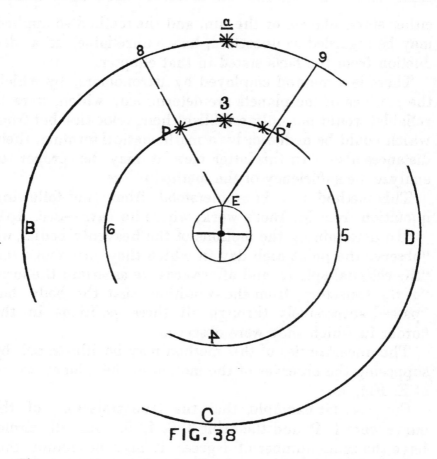

FIG. 38

For, were it moving in the lesser orbit say, for example, from P P″ yet it would appear to an observer at E to have the trajectory 8 9.   Hence, any calculation based on observation would place the planet in a larger orbit than it were in, and were the velocity or the distance determined, from such supposed trajectory would lead to an enormous deduction.

The observer would be governed by the appearance, as he would not have otherwise any correct conception of the actual distance of the planet, and to such observer it would appear as though it were amongst the stars.

That this conclusion is based on authority will appear

from the following quotation from Newcomb's Popular Astronomy.

He says, were the sun and stars visible at the same time, the sun would appear to be amongst the stars.

He thus accounts for such phenomena ; he says, this "is owing to the fact that when we look at two objects at a "great distance from us, and both on the same line of vi-"sion (as for example, a planet at P and another at 8, but "but at different distances from each other, as the planets "at P and 8 would be) we do not by mere sight recognize "this difference, hence, such two objects, (or one object at " different distances, on the same line of vision) appear at "the same distance." *

As the result of the visual insufficiency just mentioned, the observer at E would be unable to determine whether the real trajectory were P P″ or 8 9, hence could not determine either the velocity or distance of the observed planet, by such observation.

The difficulties encountered in attempting the determination of a planet's motion and consequent velocity, as also its distance, is further shown in the following quotation from Newcomb's Popular Astronomy, page 93, where it is said :

"The problem of computing the motion of each planet "under the attraction of all the others is, however, one of "such complexity, that no complete and perfect solution "has ever been found."

On page 94 it is said :

"To put the difficulty in a nut shell, the geometer can-"not strictly determine the motion of the planet until he "knows the attractions of all the other planets on it, and "he cannot determine these without knowing the *position* "of the planet, that is, without having solved his problem."

These admissions, legitimately analyzed, amount to a

* The present writer has made the parenthesis, in the above quotation.

'round about way of practically acknowledging that the distance of no planet is known.

Because if the position, that is to say, the *locus in quo* of the planet is unknown with reference to the other planets, and the positions of the other planets are unknown—the earth being classed as one of the planets—so that the geometer cannot solve the problem of their attractions—and these are the statements—this is tantamount in effect to that of a direct admission that the *distances* of the planets are unknown.

However, the distances of the planets, from the sun, are deduced from Kepler's third law, which is as follows :

"The square of the time of revolution of each planet is "proportional to the cube of its mean distance from the "sun."

M. Biot—Sec. 433—details the method by which Kepler found this law, namely :

"By comparing in a variety of instances their revolu"tions with their distances, Kepler discovered this singular "relation ; that the squares of the times of their revolu"tions are proportional to the cubes, of their mean dis"tances. This is Kepler's third law."

There can be no doubt that the times of the revolutions of the planets are known. This knowledge is obtained by observing the time of the return of each planet to the node of its orbit.

But how could Kepler have accomplished the other necessary part of this task ? namely, that of comparing the *distances* with the times of revolutions of the planets. For astronomers tell us—as before quoted— that the distances of the planets may be found after the distance of the sun is found, but *not* before.

That the sun's distance has not been found, is evident when astronomers tell us, as they do that "the distance of the sun is one of the capital problems of modern astronomy," and this distance was sought but a few years ago on the occasion of the transit of Venus over the sun's disc.

Were it maintained that Kepler's third law not only establishes the proportional, but the real distances of the planets, then one of two assertions of astronomers in regard to planetary distances must be untenable for the reason that they are contradictory of each other.

For instance, if the distance of a planet results from a proportion found between the cube of its distance, and the square of the time of its revolution, and this proportion results, and the distance of the planet is found independent of a knowledge of the *real* distance of the sun, which must have been the case with Kepler, as such distance *yet* remains a *problem*, can it be true, as asserted by astromomers, that a knowledge of the distance of the sun is *essential*, hence, prerequisite to finding the distances of the planets?

On the other hand, if a knowledge of the sun's distance be thus essential to finding the distances of the planets—and as quoted astronomers say it is—and the distance of the sun be not yet found, and it is admitted that it has not, then it will follow that Kepler's discovery of this "singular relation" cannot determine the *real* distances of the planets.

And as a corrollary, it may not be impertinent to remark in this connection that if the third law of Kepler does not find the *real* distances of the planets, how can taking the parallax on the occasion of the transit of Venus over the sun's disc, aid in supplying the deficiency in the knowledge of the sun's distance?

As the distance of the sun is unknown, how could this proportionate relation of the cube to the square have been found by Kepler?

In such case it could only have been found by squaring the time of revolution of a planet—for the time of its revolution is known—and then finding by calculation, a distance whose unit when cubed would be proportionate to the square of the time of revolution of such planet, and then to assume by reason of this "singular relation" that the distance to the sun were in accordance with such proportion.

But this would only show a proportionate relation, and not the real distance of the sun or the planets, for were the distance of the sun from a planet reduced, for instance the earth, and were the distances of all the planets from the sun diminished in *like* proportion, the *cubes* of the distances of the planets from the sun would remain proportionately the same, hence, the squares of their times of revolution would be in the same proportionate relations to the cubes of these reduced distances as they were before such reductions.

This is almost self evident, but an example may be permissable.

For example, let the cube and square of revolution of the planet Mars be calculated.

The mean distance of this planet from the sun, according to astronomical estimate, is 141 millions of miles.

On the basis of the nearest astronomical distance of the sun to the earth being 92,500,000 miles, let this distance be taken from which to find the cube of Mars' distance.

By employing 92,500,000, the astronomical distance of the sun, as a divisor of 141 millions, the estimated mean distance of Mars, we have as the quotient, 1.524, which gives the *unit* distance of Mars, which is as laid down in astronomical works of authority.

The cube of 1.524 is 3.539. The period of revolution of Mars is found, by observation, to be 1.881 years.

The square of 1.881 may be found to be 3,538, which shows, in accordance with the third law of Kepler, that the square of the time of revolution of Mars is almost in identical proportion to the cube of its unit distance.

Now suppose the distance of the sun to be reduced to 1,000,000 miles from the earth, which is in accordance with the new measurement of such distance, as explained in Chapter II, and suppose that the distance of Mars from the sun were reduced proportionately; that is to say, in the proportion that 92,500,000, the estimated astronomical distance of the sun, bears to 1,000,000, the reduced distance of the sun, according to the new measurement of such distance shown in Chapter II, and the distance of Mars from the sun after reduction—using 92,500,000 as a divisor of 141,000,000—will be found to be, in round numbers, 1,524,000 miles.

On the basis of such reduced distance, employ the same method to find the *unit* distance as was employed for that purpose before such reduction, namely, using the *reduced* distance of the sun as a divisor of the *reduced* distance of Mars, that is to say, divide 1,524,000 by 1,000,000, and the *unit* distance of Mars will be found to be as it was before such reduction, namely, 1.524.

The cube of 1.524 is, of course, as it was before such reduction, namely, 3.539, and the square of the time of its revolution will be found to be 3.538, as before.

It will be seen from the foregoing calculation that the proportion of the cube of the distance to the square of the time of revolution, is relatively the *same* for Mars after reduction of its distance, as it was before such reduction.

The writer finds, by calculation, the same identity of results with all the planets, as that found for Mars.

The same identity of proportion between the cube of the distance and the square of the time of revolution, whether the distance assumed be 2, 3, 4, 6, 6½ millions, or any other proportional number of miles, as the sun's distance, which, of course was to have been expected.

This fact shows, at least so far as in the instances just mentioned, that the proportion found between the cube of the distance and the square of the time of revolution is a "singular relation" based on an assumed distance of the sun.

If the sun be but 1,000,000 miles from the earth, the planets must be nearer the earth than the distances at present given them, in order that their attractions be in proportion to such reduced distance, and proportionately reduced mass of the sun.

To find such distances, the same method may be employed as that which found the reduced distance of the sun, explained in connection with Fig. 5, Chapter II.

To do this, let a telescope and plumb line find the zenith of the planet.

At 40° or 45° from such point, let the angular line—corresponding to the angular line A S, Fig. 5—point on the observed planet at the time it is on the zenith 40° or 45° distant.

Then pursue the same method as that which found the distance of the sun, detailed in Chapter II, in connection with Fig. 5, and thus the distance of the planet may be found, provided A S forms a measurable angle with the zenith or plumb line.

Say, for example, that the telescope and plumb line were at the equator, and the observed planet on the zenith there, and at the same time the line A S, Fig. 5, were pointed on such planet at 40°, 50°, or 60° North or South of the equator ; were such planet so distant that the line A S would be parallel with the plumb line at the equator, that is to say, were such planet at the same time vertical on the line A S and on the plumb line, the method proposed could not measure the distance of *that* planet.

———

In Chapter III an attempt has been made to show that the facts and phenomena relied on by astronomers in proof of the earth's rotation are susceptible, philosophically, of an interpretation independent of such asserted rotation.

In the present section, the attempt will be made in-tended to show that the observed phenomena in connection with the moon's motion, may be rationally explained on the theory that the earth does not rotate.

So far as the earth is involved, the sun is thought to be without progressive motion, hence the theory of eclipses of the sun and moon is that they are caused by the mo-tions of the moon and the earth.

The phenomena of eclipses of the sun and moon are at-tributed, by astronomers, to the monthly motion of the moon about the earth from West to East, combined with the motions of the earth in the same direction.

The supposed *easterly* motion of the moon is, it is said, fully confirmed by the fact that it recedes from the stars in an easterly direction, and by the additional fact that the western limb of the sun is *first* eclipsed, and that such eclipses move gradually from West to East over the face of the sun, which phenomena are regarded by astronomers as decisive evidence that the direction of the moon's mo-tion is from West to East.

However, should it be the fact that the earth is immov-able, and the sun revolves about it, as the moon and the outer planets do, and as a portion of the stars appear to do, namely, those which appear to set in the West, then the moon must move around the earth from East to West; for on the hypothesis that the earth does not rotate, an eastern direction of the moon's motion would make it im-possible for the new moon to set in the West, which would practically be in contradiction of observed phenomena, which phenomena, however, are consistent with the pres-ent astronomical system, as this result is attributed to the rotation of the earth from West to East.

Were it supposed that the earth is immovable, that the sun revolves about it, and that the stars, sun and moon revolve in a westerly direction, in accordance with the ob-served phenomena, and the moon to have a slower motion than the sun and stars, all of which must be the case if the earth does not rotate in order to account for observed phenomena, on such hypothesis, the sun and stars will present the same phenomena in their motions over the moon, as those observed at such times, and which result, according to the present system of astronomy, from an eastward motion of the moon, thus moving, apparently, over a motionless sun and motionless stars.

As an objection to this theory it may be said that it is not philosophical, because, instead of deducing the theory from the facts, the facts are imagined in order to support the theory.

To such hypothetical objection it may be replied :

1. That the theory is based on observed phenomena, namely, the immobility of the earth, and the westward motions of the stars, sun and moon, and the fact of such motions, is a deduction from what such observed phenomena legitimately imply.

2. Astronomers do not assert that there is any law of nature adverse to such hypothesis, but merely claim that the present solar system is more simple and more probable because of the supposed great distances of the sun and stars.

In any case, the supposed objection to the theory here advocated, would come with a bad grace from those maintaining the present solar system, for the reason that observed phenomena, which appear to contradict the Copernican theory, are metamorphosed into something other than what such phenomena would normally indicate.

A few instances of this illogical method may be given by way of example :

1. The earth appears motionless. It is made to rotate on its axes, and to progress about the sun.

2. The sun has the appearance of revolving about the earth, as the moon has. It is made to stand still in the heavens.

3. All the heavenly bodies appear to move westward. The Copernican theory reverses all these apparent directions of motion and turns them eastward.

4. The immobility of the stars suggests the improbability that the earth has a progressive motion.

The absence of displacement of these bodies as a result of such progressive motion is accounted for by giving the stars such vast distances that their displacement would not be observed as a result of such progressive motion ; yet the distance of no star is known.

5. The constellation Hercules is observed to move westward, and in the direction of the sun. This is accounted for on the theory that the sun progresses eastward.

But to continue the consideration of the theory herein maintained :

If the fact be in accordance with the appearance, namely, that the stars move westward, the appearance of a star East of the moon, and soon after West of it would result from the proper motion of the star, moving westward with greater velocity than that of the moon moving in the same direction.

In such case the appearance would be that of the moon moving eastward.

The appearance would be similar to that of two steamboats moving in the same direction, and one of them passes the other. At this instant, the slower—analogous to the moon—has the appearance of moving in a direction opposite to that of its real motion.

In the case of the sun's eclipse, first appearing on its western limb, and the eclipse thence moving over the sun eastward, adduced as evidence that the moon moves eastward, it will be readily perceived that the appearance would be similar were both sun and moon moving West, as both appear to do, were it also the fact that the sun really moved the faster, for in such case the western limb of the sun would first touch the eastern edge of the moon, and as it, in its more rapid western motion, covered the body of the moon, the appearance would be that of an eastward moving moon, eclipsing an immovable sun ; and the moon would have this appearance in the case of an eclipse of a star, were it the fact that both moon and star moved westward, and the star moving the faster.

Such would be the appearance for the reason that it is an observed fact that where two bodies are moving in the same direction, parallel to each other, but at a great distance from each other, the body nearer the observer—in this case the moon—always appears to be in motion, while the further body appears to be motionless ; that is to say, such is the appearance when the nearer body is *very* much nearer the observer than the further one, as is the moon in comparison with the much greater distance of the sun from the observer.

In consequence of this phenomena of vision manifested in connection with bodies moving parallel to each other, in the same direction, and at a great distance apart, the moon would appear to touch the western limb of the sun on their first apparent contact, while at the same time the sun would appear motionless.

From these appearances alone it cannot be determined which of these two adverse theories is true, as to the motions of these heavenly bodies, for the reason that the phenomena, observed on the occasion of an eclipse of the sun, or of a star, may be rationally explained consistently with either theory, on the hypothesis—and each theory is founded on hypothesis—that all these bodies move westward, and that the sun and stars are very much further from the earth, than is the moon.

If it be the fact that both the sun and the moon revolve about the earth in a westward direction, and make the circuit of the heavens daily, so to speak, there would .be an observed difference between the velocities of these bodies, the sun moving with the greater velocity, as it may be ob-

served that the moon has apparently retrograded eastward
each night on the celestial sphere, a distance which would
require about an hour's time for it to reach the position it
was in on the preceding night at the same time, did it
move westward, which would require twenty-five hours to
complete its circle about the earth, of the preceding night,
while the sun appears to complete the circle about the
earth in twenty-four hours.

In regard to the eclipses of the moon, these are caused
—as is well known—by the shadow of the earth being
projected into space, as an effect of the sun shining on the
hemisphere of the earth, opposite the shadow, in which
shadow the moon becomes enveloped, at times of its
eclipses, in revolving about the earth.

As the moon moves in an orbit about the earth, it must
become eclipsed as a result of the cause just stated, independent of the fact that it moves either East or West, and
a little reflection on the subject will satisfy the mind that
such result would occur whether the earth rotates or not,
did the sun and moon revolve in a westerly direction
about the earth, for it is the positions of the sun, earth,
and moon, with reference to each other, and to the observer, which causes the phenomena, and not how such
relative positions were brought about.

The totality of an eclipse of the moon may continue,
and usually does continue about two and one-half hours,
and including the penumbra or shading less than total ob-
scuration, it may continue, and usually does, as long, and
sometimes, somewhat longer than four hours, but about
two and one-half hours is the usual duration of a *total*
eclipse ; although this estimate may be but an approximation, because, "The shadow of the earth is but little darker
"than the region of the earth next to it. Hence, it
"is very difficult to determine the exact time when the
"moon passes from the penumbra into the shadow, and
13

"from the shadow into the penumbra ; that is when the "eclipse begins and ends." Parker's Phil, Page 395.

According to the theory herein maintained, eclipses of the moon can only take place as a consequence of the earth's shadow passing over the moon, resulting from the velocity of the sun being greater than that of the moon.

The velocity of the earth's shadow would be proportional to that of the sun's, and would be greater than that of the moon's, hence, instead of the moon entering this shadow—which is the received opinion—the more rapidly moving shadow would pass over the moon, and thus eclipse it.

For example, suppose the full moon to be on the meridian of an observer, and the sun to be at the same time on, or near, the meridian, over the earth over the hemisphere opposite the moon.

In such case, the shadow of the earth would be projected into space, in the direction of the zenith over the place of the observer. when the moon there would be vertical.

The sun being thus at the moon's antipodes, so to say, at midnight of the observer would move eastward, while shadow of the earth cast by it in the neighborhood of the moon would move westward, as a consequence of *this* eastern direction of the motion of the sun.

As the shadow, moving westward with greater velocity than that of the moon, eclipses it, the eastern limb of the moon will be eclipsed *first,* just as it is observed to be, and the shadow will pass over the moon, thus eclipsing it.

That the shadow would move westward with a greater velocity than the moon moving in the same direction, according to the theory herein maintained, may be shown. According to this theory, the moon revolves about the earth in 25 hours. The mean distance of the moon from the earth is 240,000 miles, on this basis its velocity in revolving about the earth in 25 hours will be about 57,000 miles an hour.

The velocity of the sun—such body being one million miles from the earth, according to the new measurement of its distance—moving about the earth in 24 hours will be 250,000 miles an hour.

The shadow of the earth where it would envelope and eclipse the moon, would be at the distance of the moon from the earth, namely, 240,000 miles.

If the distance of the sun from the earth, namely, one million miles, be regarded as a lever of the longer arm, the earth as the fulcrum, the shadow as the shorter arm of this lever, and the moon as the weight or resistance, on mechanical principles, of the extremity of the longer arm of the lever moved with the velocity of 250,000 miles an hour, the shorter arm of the lever, namely, the earth's shadow at the distance of the moon, namely, 240,000 miles from the earth, would move with a velocity of 60,-000 miles an hour.

The moon, 240,000 miles from the earth and revolving about the earth in twenty-five hours, would have a velocity, in round numbers, of 57,000 miles an hour.

Hence, the earth's shadow would have a velocity in excess of that of the moon equal to 3,000 miles an hour, equal to 7,500 miles in 2½ hours, during which time an eclipse would probably continue, as it would require about that time for the shadow to pass beyond the moon, and thus unveil its light to the supposed observer of its eclipse.

From the foregoing it will be observed that the phenomena of the moon's eclipses may be accounted for on the theory of the motions of the moon and sun herein advocated.

It might be supposed that heat would be generated, according to the mechanical theory of heat, did the moon revolve about the earth in twenty-five hours, through the resisting medium which occupies space, with a velocity of 57,000 miles an hour.

Yet it is a fact that the moon furnishes no evidence of having heat. This is proved by the fact that the rays the moon, concentrated on a lens, and thence directed on to a spirit thermometer. do not raise its temperature.

The want of heat of the moon may be the result of igneous action, of which it bears evidence. It is possible that its combustible matter, by the constant and long continued action of intense heat, may have been converted into an incombustible residuum, and incapable of manifesting heat with a velocity of 57,000 miles an hour; or, this velocity may have brought about this result in the long lapse of innumerable centuries, so that the matter composing the moon is no longer capable of manifesting this phenomenon.

In concluding further consideration of the theory maintained in the present section, it may be pertinent to quote M. Biot, who says:

"It is impossible by mere observation to say whether "the sun or the earth is in motion."

This amounts to a tacit admission, from very high authority, of the abstract possibility of the motion of the sun and immobility of the earth, maintained in these pages; for were such inconsistent with natural law or observed phenomena, "mere observation" would never have been invoked as a means by which the question could, possibly, have been settled.

In the present section, we present for consideration some observed natural phenomena which may aid in showing the plausability of the idea, herein maintained, that the stars revolve about the earth from East to West.

An attempt will also be made to show the insufficiency of the evidence relied on by astronomers of their vast distances, which, however, never have been measured.

In order to convey an idea of their inconceivably great distances, M. Biot, in Secs. 52 and 53, says that,

"The dimensions of the earth compared with the dis-"tances of the stars are almost infinitely small. Then, in-"deed, the visual rays, drawn from different parts of the "surface to the same star, ought to appear parallel; or, "what amounts to the same thing, the earth as seen at the "distance of the stars, would seem but a point. * * * "Since all the visual rays leading to them at the same in-"stant from *all* parts of the earth are considered as paral-"lel, it follows that the different observers would project "them upon these spheres in points exactly correspond-"ing."

It requires no argument to prove that the lines formed by the visual rays mentioned in the above quotation would not really be parallel, because it is a mathematical axiom that lines drawn from different points to the centre of the same sphere cannot be parallel, and the visual rays mentioned would be as lines so drawn.

Were such rays, or lines really parallel they could not meet in the same star, as it is also an axiom that parallel lines cannot meet.

Yet where a triangle has a very narrow base in comparison with the length of its respective sides, such sides may *appear* as though parallel to each other, and such would be the appearance, were it possible to draw lines from different parts of the earth to the same star, as may be deduced from the last quotation. Were such parallelism the only fact from which astronomers estimate the distance of the stars to be two million radii of the earth's orbit, such would not be a necessary deduction from such parallelism, as will be presently attempted to be shown.

But it does follow, as a necessary consequence of the apparent parallelism of the lines mentioned in the quotation, that parallax of the star, attempted anywhere, where such lines appear parallel, is an impossibility as "the par-

"allax of the heavenly body is the *angular* distance be-
"tween the true and apparent situation of the body."

According to this definition of parallax three prerequisites must exist before parallax of a heavenly body is possible, namely:

1. The real situation of the heavenly body.

2. Its apparent situation, different from its real situation
must be known, and

3. The *angular* distance between *two* situations.

None of which prerequisites can be known in the case
considered, as lines drawn from *any* two different places
on the earth will appear parallel, as is virtually stated to
be the fact, in the last preceding quotation from M. Biot,
hence, there are no *two* situations of the star, answering to
the requirements of the true and apparent situations of
such heavenly body, which are necessary in order to take
its parallax, and in such case there does not exist any
angular distance susceptible of measurement, and for these
reasons parallax of any star is an impossibility *per se.*

And for the same reason the distance of the stars,
namely, two million radii of the earth's orbit, is conjectural, as such distances are not pretended to be *known*
otherwise than by parallax where the diameter of the
earth's orbit is used to form the base line of the necessary
triangle, and where the same apparent parallelism of its
lines drawn towards the same star is observed as when
similar lines are drawn towards the same star, from different parts of the earth.

This apparent parallelism of lines drawn, say from opposite sides of the earth or from opposite sides of the
earth's supposed orbit towards the same star, does not
necessarily indicate that the distance of such star must be
that assigned it by astronomers, nor indeed that it is at a
great distance, in comparison with such assigned distance.

This may be shown as follows: for example, an examination of Fig. I. will show that the vertical line repre-

senting a line supposed to be drawn from the equator to the sun, and the line A S drawn at an angle to it, from 40° N to the sun *appear* to be parallel.

Yet the distance of the sun from the earth measured by means of the apparently parallel lines, is but one million miles, which distance is almost incomparably less than the distance assigned the stars, partly from the same cause, namely, the apparent parallelism of lines drawn, as is supposed, from opposite sides of the earth's orbit.

Again, were the base of a triangle one inch in length, and the lines forming its sides 100 feet long from base to apex, such lines would appear parallel to each other.

This fact shows that apparent parallelism is possible where the distance from the base of the triangle to its apex (namely, 100x12 equal 1200), does not exceed 1200 times the length of its base.

Hence, it is *possible*, in accordance with the experiment that the distance of a star *may* not exceed 1200 times the length of the base of the triangle, from whose extremities the distance of a star is attempted.

Then, were lines supposed to be drawn from opposite sides of the earth to the same star—such lines being apparently parallel, the distance of such star—by analogy to the experiment above mentioned, and as an inference from such apparent parallelism—might not exceed 1200 times the length of the base line, from the extremities of which such lines may be supposed to be drawn to the same star.

Were such base line the diameter of the earth, it would be, say 8,000 miles in length and 8000x1200 equal 9,600,000.

Hence such star, measured from the extremities of this base line, might possibly be—as indicated by the apparent parallelism of the lines supposed to be drawn to the same star—9,600,000 miles from the earth.

And even lines,—as shown by the shorter lines in Fig. 1—shorter than these might be apparently parallel, and as a result the star might be even at a less distance than that just given.

For while there is ample and indubitable evidence that the stars are more distant than the sun, such evidence does not exist that the stars are distant two million radii of the earth's orbit.

While it is true that the earth is a *point* at the distance of a star—the impossibility of finding the parallax of a star proves this—it is also true that a star is a point at the distance of the earth from it.

In this view it may be proper to consider whether or not, did the earth rotate, such rotation would give stars, in the same quarter of the heavens as that of an observed star, an appearance of revolution, in case such stars were so situated that they could not be concentric with the axis of rotation of the earth.

For example, take the stars constituting the cluster named *Ursa Major*, or the Great Bear.

This cluster appears to revolve, the higher stars from left to right, the lower ones from right to left, if they are observed through a telescope ; but astronomical telescopes apparently reverse the direction of the motions of heavenly bodies observed through them, hence this appearance indicates a motion of these stars from right to left for the higher, and from left to right, for the lower stars.

This cluster, however, is asserted to have no motion of revolution ; the appearance of such is attributed, by astronomers, to the rotation of the earth.

If the earth be a mere point—at the distance of the stars—and lines drawn from opposite sides of it meet in the same star as parallel lines how is it possible for the rotation of such point, namely, the earth, to be the *cause* of the appearance of revolution which is observed of this cluster.

In order to show, by an illustration, the improbability that the rotation of the earth is the cause of this appearance, suppose a large surface to be on a plain or prairie perpendicular to the horizon, and having luminous points at intervals all 'round the circumference of such surface, these points will represent the stars constituting the cluster named the "Great Bear" although these, as a cluster, are not placed in a circle.

Suppose a globule, say one-sixteenth of an inch, or less, in diameter to represent the earth, and that such globule were placed to the eye, and the line of vision thence directed to *one* of the luminious points on the surface mentioned. Let parallel lines be drawn from opposite sides of this globule and let it be supposed that such lines would meet in *one* of these luminous points, and at the same time be parallel with each other.

This comparatively, almost infinitely, small globule, and luminious points just mentioned may be regarded as analogous to the earth reduced by distance to a point, so that lines drawn from its opposite sides would meet in a star and yet would be parallel to each other.

Let it be supposed that while these parallel lines are pointing on one and the same star, that the globule is rotated on its axis; the following deductions may be made, namely :

That if the luminous points mentioned, had no motion of revolution, the rotation of the globule—whose parallel lines meet in one and the same star—could not in accordance with any known law of vision, or any known law of motion, give the circle of luminous points mentioned, the appearance of revolving, and conversely did it really have such appearance it could not be the result of the rotation of the globule.

It may be conceded that the vision of the observer—supposing him placed on the globule—would not be necessarily confined to the luminous point observed from the globule, but might observe the revolution of the whole circle of luminous points, but such observed phenomena would not be the result of an optical illusion caused by the rotation of the globule, for the reason that the sweep of its rotation would not extend beyond the range of the luminous point to which the supposed parallel lines were drawn ; hence, the sweep of the globule's rotation being confined to but one of the luminous points could not produce the optical illusion that all the points were in motion as an effect of the circumscribed rotation whose extremest sweep does not extend beyond the confines of a single star.

Hence, did such circle of luminous points have the appearance of revolution it would indicate a real and proper motion of the group.

This illustration may appear more clear, were it supposed, for example, that the surface mentioned as being perpendicular to the horizon were five miles in diameter, and that fifteen globes, a mile apart, and each 100 yards in diameter were placed 'round the circumference of such surface, or placed irregularly on it, and the globe, corresponding in position to the globule before mentioned, were also 100 yards in diameter, and were placed, say, one-half mile from such surface. Let it be supposed that lines were drawn from opposite sides of the latter named globe to the opposite sides of a globe on the surface mentioned, and that such lines were parallel, and at right angles to the surface to which they were drawn.

Now suppose such globe, placed one-half mile from such surface, to be rotated on a central axis of motion.

To an observer on such rotated globe, the globes on the surface mentioned, and placed a mile apart, would not, to such observer, have the appearance of revolving, for the reason before given, namely, that the circle described by such rotating body would not include the globes 'round the circumference, or irregularly placed, as the case might be, on such surface.

If this argument be legitimate, on the basis that parallel lines would meet on the same star, if drawn, say, from opposite sides of the earth, the argument deduced from the examples just given, will apply to the cluster, of stars mentioned, and if so the rotation of the earth will not account for the appearance of revolution of the cluster; hence, the conclusion is reached, that such appearance of revolution is real, and that the cluster of stars constituting the "Great Bear" do revolve, and that the direction of such motion is from East to West.

From what has been here said the following deductions are made :

1. If the earth, by its distance from the stars, be a mere point in space, the rotation of such point will not explain the apparent revolution of the stars just before mentioned, hence, the apparent revolution of these may be rationally regarded, from this stand point as a consequence of their real revolution.

2. If the earth has a real magnitude at the distances of the stars, and its own rotation within a circle of less than 25,000 miles, namely, its own circumference, causes the appearance of the revolution of the cluster of stars mentioned, it appears almost incredible, that did the earth progress in an orbit of 570 millions of miles in circumference—and this is the assigned dimensions of the orbit—that no star in the heavens should appear to be displaced as an effect of such revolution, and this is proved, and admitted to be the fact, as no star shows any parallax as a result of this supposed revolution of the earth in its orbit.

There is another, real or apparent, motion of some stars which is attempted to be accounted for as an apparent motion by M. Biot in the following quotation—Sec. 4 :

In speaking of the sun, he says : "If we observe this lu-"minary near the time of its setting, for several days in "succession, and when it is below the horizon, examine "the stars which follow it, and which set immediately

"after it, they are easily distinguished by the figures un-
"der which they appear in the heavens ; after several
"days they are too faint to be seen. There are other stars
"which follow the sun, and set immediately after it.

"These same stars on the preceding days did not set
"till sometime after the sun ; the sun has therefore ad-
"vanced toward them from *West* to East, in a direction
"contrary to the diurnal motion.

"If we examine the heavens in the morning, sometime
"before the rising of the sun, the appearances will be the
"reverse of these, the stars which rise to-day at the same
"time, or nearly the same time with the sun, after several
"days will rise sometimes before it. They will appear to
"have receded in the heavens from East to West, or which
"is the same thing, the sun will have removed from them
"in the direction from West to East. For it is more sim-
"ple to suppose the sun to have a proper motion, than to
"suppose that all the stars have, with relation to the sun,
"a common motion.

"By this proper motion the sun seems to describe the
"whole circle of the heavens, from West *to* East."

The enquiry which suggests itself is, has this eastern
direction of the motion of the sun any other foundation
than the simplicity of the supposition that it has such
motion ; for it will be seen from the foregoing quotation
from M. Biot, that with him it rests on this basis.

There is *but* one other reason given for it, so far as the
writer has been able to find, and this is shown in the following quotation from Herschel's Astronomy, title, "Fixed
Stars," wherein Sir John Herschel gives the opinion of
Sir William Herschel on this point. He says :

"A general recess could be perceived in the principal
"stars, from that point occupied by the star Hercules,
"toward a point diametrically opposite. This general ten-
"dency was referred by him to a motion of the solar system in an opposite direction." That is to say, certain

stars appeared to be moving westward, and this phenomena he attributed to the motion of the sun and whole solar system in an eastern direction.

M. Biot—Sec. 161—says, in speaking of this supposed eastern motion of the sun :

"The motion of this body from West to East, considered "as parallel to the equator, takes place in a regular man- "ner, though not uniformly, and in the course of a year."

From the former quotation from M. Biot—which nothing said by Herschel contradicts—the asserted eastern motion of the sun has no foundation beyond conjecture.

There does not appear any reason for such conjecture, other than that if the stars move from East to West of their own proper motion, such motion *pro tanto*, contradicts the Copernican system of astromomy, which maintains that the stars have no progressive motions, but only an appearance of such caused by the rotation of the earth.

Did such eastern motion of the sun and the whole solar system exist, the earth would be included in the general eastern drift, and the sun could revolve about the earth as though it were immovable, for its motion, in common with that of the sun, would have no other effect on the sun in its revolution about the earth, than would be observed were the latter body immovable. A little reflection will satisfy the mind of the truth of this.

In order to show that there is nothing incongruous in the opinion that the proper motion of those stars westward —attributed to a real eastward motion of the sun—is the real cause of this observed phenomena, we quote from Newcomb's Popular Astronomy, part iv, previously quoted in another connection. He says :

"The stars in *all* parts of the heavens, move in *all* di- "rections with *all* sorts of velocities."

Then it would appear plausible that if the stars "move in *all* directions," that it is possible that this apparent motion westward of those stars observed in connection with the sun and attributed to the eastward direction of the motion of this luminary, may *really* move westward as they appear to do.

M. Biot does not fix on any precise period of the year at which the phenomena he mentions may be observed. The inference, then, would be that they may be observed at any time.

But if such should not be the case, and the phenomena he mentions—supposed by him to be caused by the eastern motion of the sun—as taking place at sunset, occur between December and June, and those occurring at sunrise should take place at sunrise between June and December, the phenomena he mentions would result from the observed apparent motions of the sun.

For example, each day becomes longer after the solstice in December until the time of the solstice in June, hence, between those periods, the sun, each day, would be a little longer time above the horizon than on the preceding day, and, as a consequence of this fact, the stars which did not set on the preceding day "till some considerable time after the sun," will now by reason of the day becoming longer, "set immediately after it."

These phenomema would be the result of the apparent *northern* direction of the sun's motion, at the time of the year when it appears to move North towards the tropic of Cancer.

Were the phenomena mentioned by M. Biot as taking place about about sunrise, observed between the summer and the winter solstice, the sun then being on its way to the tropic of Capricorn, these phenomena could be accounted for, as a result of the sun's apparent motion in *that* direction, because during the time the sun appears to move in the direction of Capricorn, it rises *later* each day, "and the stars which rise to-day at the same time, or "nearly the same time, with the sun, after several days "will rise somtimes *before* it," may be a mere appearance, not caused by an eastern motion of the sun, but be the result of the sun rising *later* each day, which would give these stars the *appearance* of rising *earlier* each day, which is the phenomena observed.

The fact that a star is not magnified when observed through a telescope is regarded, by astronomers, as evidence that the stars are at very great distances.

The reason for this conclusion is that as the sun, moon, and planets are greatly magnified when observed through astronomical telescopes, and the stars are not, that, therefore, the stars must be at much greater distances than these bodies.

This conclusion ignores the fact that such absence of magnitude may be caused by the insufficiency of the telescope as a proper visual agency for observing the stars; also appears to take it for granted that such want of magnitude cannot result from any other cause than the very great distances of the stars.

In point of fact, however, some stars are magnified when observed through such telescopes, for in Newcomb's Astronomy, page 119, it is said :

"If the eye-piece is pushed in, so that the star is seen, "not as a point, but as a small disc, the centre of the disc "will be green or yellow," &c.

Pushing in the eye-piece makes the telescope a new instrument, *pro tanto*, for observing the stars, for by means of this change, that which was before seen as a mere point, thus becomes magnified to the dimensions of a "small disc."

This change in results, by means of a virtual change of instrument, makes an inference plausible that the supposed great distances of the stars may not be the cause of their not being magnified, but that such may result from want of adaptation of the instrument, namely, by its not getting the focal distance of the observed star.

It is true that Prof. Newcomb attempts to explain the cause of this change in the appearance of the observed star, for the purpose, perhaps, of showing that this disc appearance does not amount to a real magnitude of such star, for he says, on page 142 :

"It is true that a bright star in the telescope sometimes "appears to have a perceptible disc, but this is owing to "various imperfections in the image, having their origin "in the air, the instrument, and the eye, all of which have "the effect of slightly scattering a portion of the light "which comes from the star."

Whether the causes assigned be the true ones or not, the result remains, namely, that a star is sometimes magnified in the telescope, which fact negatives the assertion that the stars are not magnified in the telescope, because of their great distances.

The general failure to magnify the stars observed through the telescope may be caused—

1. By want of the proper focus of the instrument.

2. From want of real magnitude of the star.

3. From the dimness of the light of the star *per se* and independent of its real or supposed distance.

Every lens should be constructed with reference to the size of the object to be observed through it, the distance of such object, and the intensity or quantity of the light by which the object becomes visible.

The necessity for a consideration of these facts may be noticed in connection with the ordinary spectacle glasses. Such glasses as are constructed for short-sighted persons, magnify objects only at ranges precisely adapted to the eyes of such persons, such glasses being constructed somewhat differently from those used by persons having defects of vision other than this.

Some glasses are constructed to magnify small objects— ordinary-sized alphabetical letters, for example,—at short distances from the eye, and will not magnify them at any other.

In such glasses, if the letters be brought quite near the eye, or be removed beyond the focal distance from it, they become, in some instances, obscured, and in others they are, apparently, entirely obliterated.

As the optical laws governing, in general, the construction of the lens,, apply to the construction of the lens of the telescope, similar phenomena arising from like causes, may be expected to result to this instrument, when objects of different and inapplicable sizes, or having greatly differing distances, or attempted to be made visible or greatly magnified by light of little intensity or quantity, are observed through them.

All, or any of which causes might result in the object not being observed at the proper focal distance of such object.

If this conclusion be well founded, the almost total absence of magnitude observed of the stars may possibly result from some, and, in many cases from all, these causes.

And while the telescope may have the focal range of the sun, moon, and planets, yet it may not have it for the stars, which insufficiency may possibly be caused by want of adaptation of the lens of the telescope, from the dimness of the star's light, or from want of its sufficiently great magnitude, or because the stars are at greater distances than the sun or planets.

If the failure to magnify the stars should be caused by the defects of the telescope, such defects may be possibly remedied by future discovery, for the telescope is not a perfect instrument, hence, cannot be supposed to be beyond improvement, and especially so when it is considered that it is of modern construction, in comparison with the antiquity of the usual capital appliances in aid of the arts and sciences, generally, as it dates back to so modern a period as to the early part of the 17th century.

History teaches that discoveries are often made which have given such an impetus to knowledge of the matters to which such discoveries relate, as to astonish even those best versed in such matters, and judging by these analogies recorded in history, such may ultimately result to the telescope.

The quantity and intensity of light emanating from an observed object, has, of course, much to do with the distinctness of its visibility, as also with the apparent size of such object.

This is so with reference to the stars, as a star of the first magnitude is so termed because it belongs to the class of the brightest.

As an illustration of this, as also showing the effect of the action of light, Prof. Newcomb, page 141, says :

"The diameter of the pupil of the eye being about one-
"fifth of an inch, as much light from the star as falls on a
"circle with this diameter is brought to a focus on the re-
"tina, and unless the quantity of light is sufficient to be
"perceptible, the star will not be seen. Now, we may
"liken the telescope to a 'Cyclopean eye,' of which the
"object glass is the pupil," &c.

Hence, whether a star be even perceptible, or not, depends on the quantity of its light. If the quantity be small, the star will appear dim, and if it be very dim, the star will be barely visible, and from this cause, might indeed be, supposably, invisible.

From the mere dimness of a star, it cannot be implied that it is therefore at a great distance, as dimness is not a necessary quality or condition appertaining to *great* distances, for the reason that it is often found in connection with an observed object, which is known to be at a short distance, for example, the dim light emanating from a tallow candle in a chamber adjacent.

As this is true, it would seem to follow that the absence of apparent magnitude of the stars may result from dimness of the star *per se,* hence, without regard to their distances.

The fact that the telescope does not magnify the stars, may possibly be because—

1. It has not the focus suitable to the observed star, as a consequence, perhaps, of too great a distance of a star in one instance. or of too short a distance in another.

2. For want of sufficient magnitude of the observed star.

3. From insufficiency of the light emanating from it, to that extent that the "Cyclopean eye" may not reflect it with sufficient intensity from its retina, so to speak.

Hence, the inference of astronomers that the stars are suns for other systems, and at such great distances as to be immeasurable, and to appear merely as glittering points in the telescope, may not be a conclusion justified from such premise.

In concluding the present work a philosophic apothegm of Sir Isaac Newton may be quoted, as showing the popular method of investigating truth, he says:

"The main business of natural philosophy is to argue *"from* phenomena without feigning hypotheses, and de-"duce causes from effects."

The founder of the Copernican system, not only does not argue from phenomena but he feighs hypotheses contradictory of phenomena, in *every* instance in which phenomena are adverse to the system he advocates,—and all observed natural phenomena having any connection with the system are adverse to it ;—his followers going even to the extent of feigning the hypothesis that the sun revolves from West to East, when the observed phenomena is the reverse of this.

And the advocates of the system do this on grounds which would be excluded in any Court of law, not, chiefly that they were inadmissible as incompetent evidence, but rather because they were not evidence at all.

WS - #0056 - 071223 - C0 - 229/152/12 - PB - 9781330075722 - Gloss Lamination